Lecture Notes in Statistics **145**

Edited by P. Bickel, P. Diggle, S. Fienberg, K. Krickeberg, I. Olkin, N. Wermuth, S. Zeger

Springer
New York
Berlin
Heidelberg
Barcelona
Hong Kong
London
Milan
Paris
Singapore
Tokyo

James H. Matis
Thomas R. Kiffe

Stochastic Population Models

A Compartmental Perspective

Springer

James H. Matis
Department of Statistics
Texas A&M University
College Station, TX 77843-3143
USA
matis@stat.tamu.edu

Thomas R. Kiffe
Department of Mathematics
Texas A&M University
College Station, TX 77843-3368
USA
tkiffe@math.tamu.edu

Library of Congress Cataloging-in-Publication Data

Matis, James H.
Stochastic population models : a compartmental perspective / James H. Matis, Thomas R. Kiffe.
p. cm. -- (Lecture notes in statistics ; 145)
Includes bibliographical references and index.
ISBN 0-387-98657-X (soft cover : alk. paper)
1. Population biology--Mathematical models. 2. Stochastic processes--Mathematical models. I. Kiffe, Thomas. II. Title. III. Lecture notes in statistics (Springer-Verlag) ; v. 145.

QH352 .M383 2000
577.8'8'015118--dc21

00-030462

CIP data available.
Printed on acid-free paper.

Camera-ready copy provided by the authors.
Printed and bound by Sheridan Books, Ann Arbor, MI.
Printed in the United States of America.

9 8 7 6 5 4 3 2 1

ISBN 0-387-98657-X Springer-Verlag New York Berlin Heidelberg SPIN 10697401

To Jeanette, Christine, Marianne, Barbara, and Timothy.

JHM

To Jacqueline, Tasha, Belle, Alex, Rocky, and Bunny.

TRK

Preface

This monograph has been heavily influenced by two books. One is Renshaw's [82] work on modeling biological populations in space and time. It was published as we were busily engaged in modeling African bee dispersal, and provided strong affirmation for the stochastic basis for our ecological modeling efforts. The other is the third edition of Jacquez' [28] classic book on compartmental analysis. He reviews stochastic compartmental analysis and utilizes generating functions in this edition to derive many useful results. We interpreted Jacquez' use of generating functions as a message that the day had come for modeling practioners to consider using this powerful approach as a model-building tool. We were inspired by the idea of using generating functions and related methods for two purposes. The first is to integrate seamlessly our previous research centering in stochastic compartmental modeling with our more recent research focusing on stochastic population modeling. The second, related purpose is to present some key research results of practical application in a natural, user-friendly way to the large user communities of compartmental and biological population modelers.

One general goal of this monograph is to make a case for the practical utility of the various stochastic population models. In accordance with this objective, we have chosen to illustrate the various stochastic models, using four primary applications described in Chapter 2. In so doing, this monograph is based largely on our own published work. We take this approach because we are obviously very familiar and can speak with some authority about the practical motivation, utility and shortcomings of each application. We hope that the modeling user communities will find these examples and ensuing discussion useful.

The monograph presents an approach to analyzing the stochastic models which is undoubtedly new to most readers. Our attempt is to keep the general mathematical level equivalent to that in Jacquez [28] and Renshaw [82]. The theoretical development is outlined in Chapters 3 and 9, and the methods are then applied to the models in the body of the subsequent chapters.

Specialized methods outside the scope of those presented in Chapters 3 and 9, as well as detailed extensions, are covered in the chapter appendices.

We acknowledge with pleasure and great gratitude the past and present contributions of our many colleagues and collaborators. John Jacquez, Eric Renshaw and Qi Zheng gave valuable comments and encouragement at early stages of this work. Among our other esteemed colleagues are Mel Carter, H. O. Hartley and Aldo Rescigno in compartmental analysis; David Allen, Tom Wehrly and P. R. Parthasarathy in stochastic modeling and statistical theory; and W. C. Ellis, Bill Rubink, Bill Grant, Bernie Patten, Gard Otis and Rob Hengeveld in scientific applications. Most of their names occur often in the references, where their contributions are apparent. We are most grateful to them, and to our many students over the years. Finally, this monograph would not have been possible without the skill and dedicated efforts of Yvonne Clark, our technical secretary and advisor, and we thank her for her many years of service. At the same time, we assume complete responsibility for all possible errors and shortcomings.

Table of Contents

Part I Introduction

Part II Models for a Single Population

Part I Introduction

1

Overview of Models

1.1 Modeling Objectives

Many types of mathematical models have been proposed in the literature for describing biological populations [6, 19, 28, 69, 80, 82]. One might argue that the wide variety of such models is a natural consequence of the great diversity of the overall objectives for such models. We start by discussing briefly the general modeling objectives upon which this monograph is based.

The general objective of research outlined in this monograph is to develop models which not only describe but also explain, at least in part, changes in population size(s). Accordingly, the models seek to understand the basic kinetic structure of the process which generates the observed data. Therefore, the models may be classified as largely *mechanistic* (or "explanatory"), as opposed to *empirical* (or "correlative") [80]. Most of the models of interest are sets of differential equations, where the underlying parameters are related to the assumed kinetics of the process.

Ideally, such mechanistic, kinetic models could be used to analyze population data, to make statistical inferences relating to population size, and ultimately to predict, or even help manage, population size. Whenever practical, analytical solutions are sought for these mathematical models, and when that is not available, numerical solutions are investigated. In either case, as a rule, the models in this monograph are required to be tractable for subsequent statistical data analysis. This requirement limits the types of models under consideration, for example it rules out the large-scale ecological computer simulation models.

This monograph focuses on *stochastic*, as opposed to *deterministic*, models. In our view, recent advances in computer software have greatly enhanced the practical utility of using certain stochastic models to describe biological population behavior. The monograph introduces some data sets relating to biological populations, and proposes various mechanistic models to describe the data. In general, a deterministic formulation of the model is first used to analyze a given data set. This is followed by an analogous stochastic formulation, which is also used for the analysis of data. The advantages, and challenges, of using the stochastic approach are discussed for each mechanistic model.

It is natural to question the completeness, even correctness, of any given model. Our viewpoint is stated well by Box [10] in his widely quoted statement, "All models are wrong, but some are useful." He later elaborated on this, adding "Models, of course, are never true, but fortunately it is only necessary that they be useful." [11]. Kac [32] uses a delightful analogy to make this point. He states that "Models are caricatures of reality, but if they are good models, then like good caricatures, they portray, though perhaps in distorted manner, some of the features of the real world."

The purpose of this monograph, then, is to present some stochastic "caricatures" for the description and statistical analysis of the kinetic behavior of real-world systems. Some of these models, e.g. the calcium clearance models in Chapter 10 Appendix, are quite sophisticated and incorporate detailed mechanistic concepts. Others are clearly rudimentary and are still in a preliminary stage. Yet we have found these stochastic models, whether sophisticated or still arguably naive, to be useful in general, and will make a case in this monograph for their relevance and utility in the analysis of real-world kinetic data.

1.2 Structure of Monograph

The monograph is divided into three major parts. Part I includes a detailed discussion of four data sets which are used throughout the monograph. Two of the applications relate to the dispersal, including growth and spread, of natural populations. One concerns the spread of the Africanized honey bees (AHB) in North America, based on parameter values which are given in the literature. The other describes the dispersal of muskrat populations in the Netherlands, for which parameter values are estimated from data. The two remaining applications relate to classical physiological modeling, one being the bioaccumulation of mercury in fish and the other the clearance of labeled calcium from an adult woman.

Part II presents various stochastic birth-death (BD) models for describing the growth of single populations. The linear immigration-death model, which has been widely developed in a compartmental context, is considered first followed by the linear birth-immigration-death model. These classic models are reviewed in Chapters 4 and 5 to establish a stochatic foundation for the nonlinear kinetic models in Chapters 6 and 7. The BD models have been widely studied in population modeling, as reviewed among others in [6, 23, 73, 82]. This monograph is restricted to a more narrow class of models than these other books. However our general approach of analyzing the stochastic behavior through cumulant functions, as outlined in Chapter 3, has not been used extensively in these previous texts. This monograph exploits the cumulant approach to develop new insights, especially into the nonlinear population models.

Part III develops the use of multidimensional birth-death-migration (BDM) models for describing biological spread. For example, the BDM processes are used for spatial modeling through a stochastic compartmental model construct, where the compartments represent geographical areas. Linear stochastic compartmental analysis without births has a long tradition of and readily available tools for real world applications, particularly of models being fitted to experimental data [28]. This monograph extends the development of stochastic linear multicompartment modeling by presenting the concepts of mean residence times and of non-Markovian (or age-dependent) kinetic models. Both of these concepts are based on a stochastic model formulation, and they illustrate useful contributions to practical methodology developed from the stochastic model.

Our general cumulant approach for multidimensional processes is outlined in Chapter 9. It is then applied to linear models without births, i.e. to a traditional compartmental system (Chapters 10 and 11), to linear models with births (Chapter 12), and to nonlinear models with births (Chapter 13). The monograph concludes with a simple illustration of the generality of the cumulant approach, by applying in Chapter 14 the methodology to a simple host-parasite model. In brief, Part III presents some stochastic multidimensional models and makes a case that they too are useful caricatures for analyzing data.

2

Some Applications

2.1 Introduction

Each of the subsequent chapters contains one or more simple examples to illustrate the concepts developed in the chapter. In addition to these simple examples, four applications which we have published in the recent literature will be used multiple times, in two or more chapters. Two or more stochastic models will be proposed to describe each of these four data sets, and a comparative analysis will help to clarify the relative advantages and disadvantages of the various models. A description of these four key applications follows.

2.2 Application to Invasion of Africanized Honey Bee

The rapid and widespread colonization by the Africanized honey bee (AHB) of much of South America and all of Central America has been called one of the most remarkable biological events of this century [86, p. 95]. From the escape of 26 reproductive swarms of AHB in Brazil in 1956, the population has grown relentlessly to many millions of colonies occupying over 20 million km^2 as illustrated in Figure 2.1. [95]. A review of the scientific literature on the AHB is given in [92].

One question of interest is the movement of the leading edge of this invasion, specifically in predicting the arrival times of the AHB to given locations. Methodology based on the gamma distribution for point and interval predictions of arrival times has been developed. The methods have proven accurate for predicting AHB arrival in Texas along the Atlantic coast of Mexico [58], and also AHB arrival years later in California along the Pacific coast [87]. However, in recent years the movement of the leading edge has slowed down considerably in the U.S. [33], probably due primarily to parasitism by the Varroa mite.

The second, related question of interest concerns the increase in popula-

Figure 2.1. Range expansion of Africanized honey bee populations.

tion size of AHB at a given site after its arrival. The AHB is classified as an "r-strategist which (by definition) discover habitat quickly, reproduce rapidly and use up the resources before the habitat disappears, and disperse in search of other new habitats as the existing one becomes inhospitable". [86, p. 104]. According to this theoretical characterization, the AHB population is expected to grow rapidly up until just before the carrying capacity is reached. This is an important property in postulating subsequently the nonlinear dynamics associated with AHB population growth.

Detailed empirical data on population growth were gathered by Otis [75, 76], who observed the population dynamics of the AHB in French Guiana daily for over ten months. These data are arguably among the best ever collected on the dynamics of any social insect. The variables he recorded include the observed mortality rate, the time intervals between reproductive events (i.e. swarming episodes), and the number of swarms per swarming episode.

Otis also developed a detailed simulation model based on daily time increments which incorporates the observed changes in population size due to colony births and deaths. Transient distributions of population size, $X(t)$, were obtained empirically assuming an initial size of 1, i.e. $X(0) = 1$, from 1000 replicates of the simulation process. As an illustration, Figure 2.2 represents the distribution of population size, i.e. number of colonies, at the end of ten months of elapsed time [54]. Means of such transient distributions could be obtained and plotted as a function of time to yield an empirical mean value function [54].

An overall important objective of this monograph is to illustrate stochastic models which could describe population size data, such as in Figure 2.2. A sequence of models will be considered. Part II starts with relatively simple single population models which often predict well though they may lack biological realism. Linear kinetic models are proposed in Chapters 4 and 5 for initial colonization periods [54], and the nonlinear kinetic models are developed in Chapters 6 and 7 for long-term and equilibrium population sizes [45, 56].

Part III considers multipopulation models which may greatly enhance the biological realism. Chapters 12 and 13 develop AHB population spread models which combine historical migration parameters with site specific information. Other features, such as more realistic distributions of time intervals between swarmings (i.e. "births") are also added. Chapter 14 introduces a rudimentary host-parasite model to describe the dynamic interactions between the AHB and the Varroa mite populations.

This AHB modeling project has been recognized as having great utility in mathematical education. The bee has always been a fascinating social insect for students to study, and the recent Africanization of bee colonies in North America has created keen public interest in African bee spread in the United States. Velez [98] mentions this research on AHB modeling as being a particularly useful example for integrating mathematical research and education. Other classroom models associated with AHB dispersal have also been developed (e.g. [42]) and at the time of this writing are being tested in the classroom [25]. In summary, the population modeling of AHB dispersal is a rich example which suggests new models for mathematical research, and which also illustrates how such mathematical modeling may be integrated with mathematical education for classroom useage.

2.3 Application to Muskrat Spread in the Netherlands

Data on the spread of muskrat, *Ondatra zibethicus*, populations have played a visible role in illustrating population growth models. Skellam's [89] classic diffusion model is illustrated using data from [17] on the muskrat invasion in Europe from 1905 to 1927. Also, the population dynamics models in [96, 97] have been applied to describe muskrat spread through western and northern Europe between 1930 and 1960.

Figure 2.2. Distribution (0-truncated) of the number of bee colonies at the end of $t = 10$ months from Otis' simulation model.

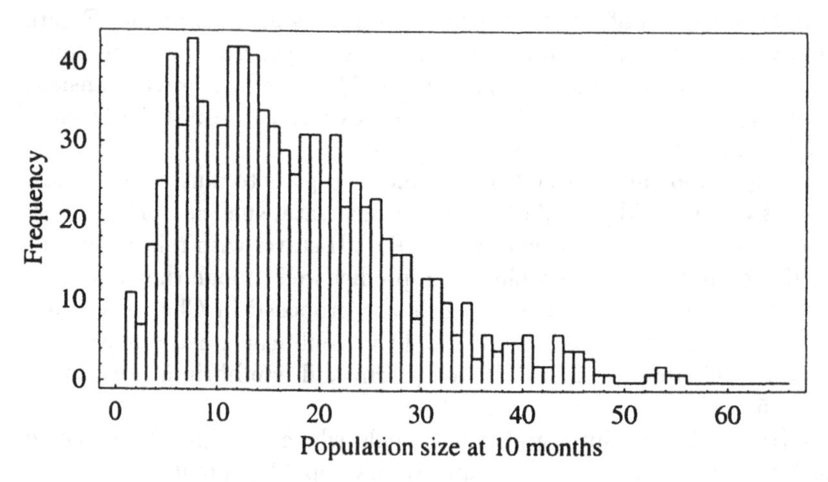

This monograph will make extensive use of data on the annual catch of muskrats in 11 Dutch provinces between 1960 and 1991 which are described in Hengeveld [26]. We assume for present simplicity that the observed muskrat catch is a measure of relative abundance of muskrats in the given province. Table 2.1 lists the province names and assigned code numbers, and Figure 2.3 locates them on a schematic map. The complete data set, from the Commission for Muskrat Eradication in the Netherlands [13], is listed in Table 2.2.

The individual data sets for the 11 provinces have some common qualitative characteristics. Each set may be divided into three periods, namely a preintroduction period, an initial population growth period, and a quasi-equilibrium period. The preintroduction period extends from 1960 to the time of first muskrat capture in the province. The growth period has rapid, exponential-type growth. The quasi-equilibrium period consists of the elapsed time after the initial growth, during which the catch usually fluctuates for a year or two and then gradually stabilizes at some reduced value.

Figure 2.4 illustrates the observed muskrat harvest data in the eleven provinces during their respective growth periods. The linear growth models are fitted in Chapter 5 of Part II to these growth period data for four eastern provinces. Nonlinear growth models, including the logistic growth model previously investigated in [26], are also used in Chapters 6 and 7 to describe individual data sets. Multi-population models are fitted in Part III to data from several provinces simultaneously, in order to estimate s

Table 2.1. Names, code numbers and abbreviations for provinces in the Netherlands.

Name	Code Number	Abbreviation
Noordbrabant	1	noordbr
Zeeland	2	zeel
Limburg	3	limb
Zuidholland	4	zuidh
Gelderland	5	gelderl
Utrecht	6	utr
Noordholland	7	noordh
Overijssel	8	overijl
Drenthe	9	dr
Groningen	10	gron
Friesland	11	friesl

migration rates and to test hypotheses concerning concurrent (adjusted) birth rates. Chapters 12 and 13 illustrate, respectively, linear and nonlinear multicompartment models to describe muskrat dispersal.

2.4 Application to Bioaccumulation of Mercury in Fish

Newman and Doubet [72] describe an experiment in which 21 mosquito fish, *Gambusia affinis*, of various size were exposed individually over a 6-day period to water contaminated with a radioactive mercury (Hg) isotope. Let V denote the size (in units of gm dry wt) of a fish and $X(t)$ its mercury bioaccumulation (in units of μg Hg) at time t. The mercury concentration, $c(t) = X(t)/V$ (in units of μg Hg/gm dry wt), was recorded daily for each fish separately using a Geiger counter. Table 2.3 lists the data for three fish of relatively large size.

This example represents a more standard application of compartmental modeling to describe a physiological system, similar to numerous relevant examples in [3, 22, 28, 85]. The overall objective in this application is to formulate a mercury transport model. It is suggested [91] that even these tiny mosquito fish store mercury in their tissue. Though there are only six data points per fish, the hypothesis of a conceptual mercury storage com-

Table 2.2. Numbers of muskrats caught annually in 11 provinces in the Netherlands, 1960–1991.

Year	Noordbr	Zeel	Limb	Zuidk	Gelderl
1960	351	42	3	0	0
1961	489	202	5	0	0
1962	1209	368	1	0	0
1963	2614	592	1	0	0
1964	4777	496	8	0	0
1965	5154	290	14	0	0
1966	8515	725	14	0	0
1967	12449	1865	10	0	0
1968	8808	1864	214	3	1
1969	12738	2688	111	2	17
1970	10628	3062	390	4	29
1971	14694	4163	634	2	461
1972	17266	4635	432	8	816
1973	23462	8005	447	52	1463
1974	30405	13880	261	261	2615
1975	41435	13066	1555	208	5024
1976	34229	8847	1509	260	6158
1977	34142	4206	1731	389	10527
1978	42737	3382	2266	798	18128
1979	38908	3630	3025	1516	18056
1980	25491	4393	2799	1541	26260
1981	24273	5619	5487	3232	40050
1982	26307	5802	6876	8315	60650
1983	26847	5733	11827	35920	60519
1984	21809	5952	8938	42654	47114
1985	30481	8456	8960	35384	48997
1986	38403	12431	8698	43288	32516
1987	30575	17096	9031	45289	32278
1988	26832	18742	13243	58508	46691
1989	21601	18279	13962	47500	47341
1990	16581	17597	12759	67437	45331
1991	12754	13031	10058	96644	50477

Table 2.2. (Continued) Numbers of muskrats caught annually in 11 provinces in the Netherlands, 1960–1991.

Year	Utr	Noordh	Overijl	Dr	Gron	Friesl
1960	0	0	0	0	0	0
1961	0	0	0	0	0	0
1962	0	0	0	0	0	0
1963	0	0	0	0	0	0
1964	0	0	0	0	0	0
1965	0	0	0	0	0	0
1966	0	0	0	0	0	0
1967	0	0	0	0	0	0
1968	0	0	1	0	0	0
1969	0	0	4	0	0	0
1970	0	0	44	1	1	0
1971	0	0	674	111	534	1
1972	0	0	791	341	452	1
1973	0	0	1072	821	58	0
1974	23	0	1295	2153	1822	1
1975	16	1	2195	2995	3512	3
1976	50	1	3023	3617	4605	23
1977	125	2	5292	8609	10024	157
1978	236	64	8107	12631	15317	150
1979	403	145	11505	17604	17898	483
1980	552	247	16277	15296	21101	857
1981	1354	588	18449	13140	26468	2555
1982	1243	673	18830	17412	24343	5897
1983	5515	513	18830	14576	19091	15253
1984	10598	453	17532	12697	18495	24643
1985	21704	587	17236	10170	16463	31908
1986	27184	1083	13899	12124	18566	40775
1987	55411	1074	14567	9476	15463	42447
1988	63581	732	16999	10617	22003	54170
1989	62000	673	17889	13362	28855	58712
1990	71000	970	17084	14570	33527	68178
1991	93039	1730	17580	14037	37420	84147

Figure 2.3. Map of the Netherlands with 11 provinces, namely 1 Noord-brabant; 2 Zeeland; 3 Limburg; 4 Zuidholland; 5 Gelderland; 6 Utrecht; 7 Noordholland; 8 Overijssel; 9 Drenthe; 10 Groningen; and 11 Friesland.

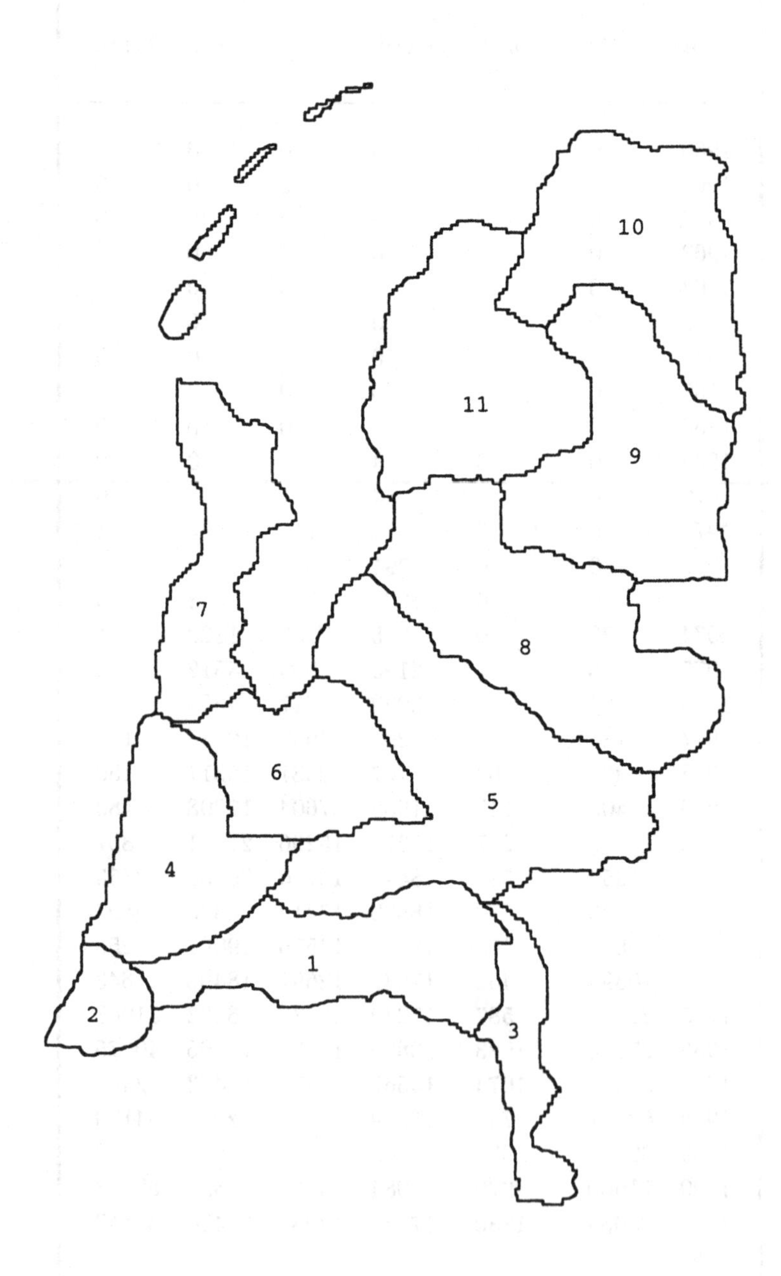

Figure 2.4. Observed muskrat catch during the initial population growth period for each of the 11 provinces

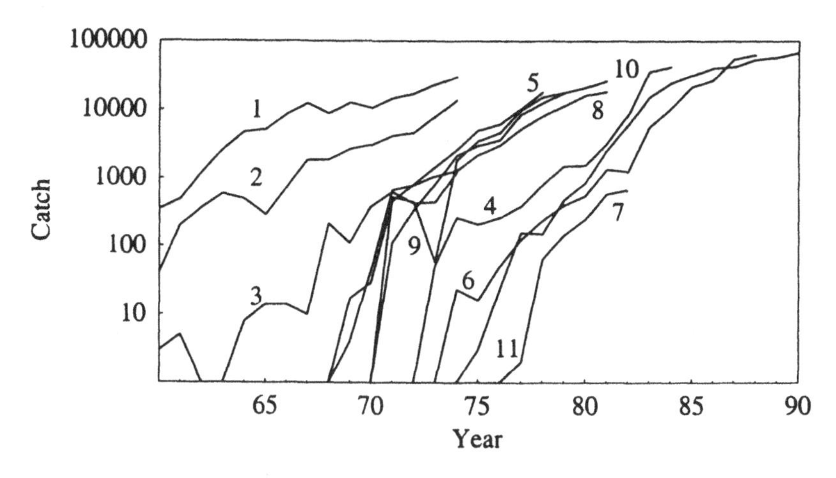

partment is examined by fitting the data first to a (homogeneous) one-compartment model in Chapter 5 and then to a two-compartment model in Chapter 8. The implications of a second (i.e. mercury-storage) compartment are then explored for this sample of fish. The complete study with results for all 21 fish is given in [57].

Table 2.3. Observed concentration over time of mercury in three fish.

Time	Concentration		
	Fish 1	Fish 5	Fish 16
0	0.000	0.000	0.000
1	0.246	0.287	0.481
2	0.354	0.450	0.669
3	0.403	0.553	0.793
4	0.461	0.653	0.893
6	0.498	0.760	0.956

2.5 An Application to Human Calcium Kinetics

As another illustration of a classical application of compartmental analysis, we consider also the excellent data set on the calcium concentration-time curve in an adult woman, considered in Weiss *et al.* [100]. The data are listed in Table 2.4 and illustrated in Figure 2.5. A commonly used compartmental model for calcium kinetics has three conceptual compartments, representing respectively the calcium in the plasma, "soft tissue", and "hard tissue" including the bone. Labelled calcium is introduced as a bolus into the plasma compartment, compartment 1, at time denoted $t = 0$. The concentration of remaining labelled calcium in the plasma is then observed at various elapsed time intervals, initially more intensively but shortly thereafter every 8 hours. This outstanding data set covers a period of nearly 600 hours.

These data are first fitted with a standard multicompartment deterministic model in Chapter 8. They are then analyzed using a linear stochastic model in Chapter 10, in order to demonstrate the concepts of mean residence times. A third analysis, which extends the non-Markovian model in [100], is given in the Appendix to Chapter 10 and is discussed in [51].

Figure 2.5. Observed clearance data of labelled calcium from plasma of an adult woman.

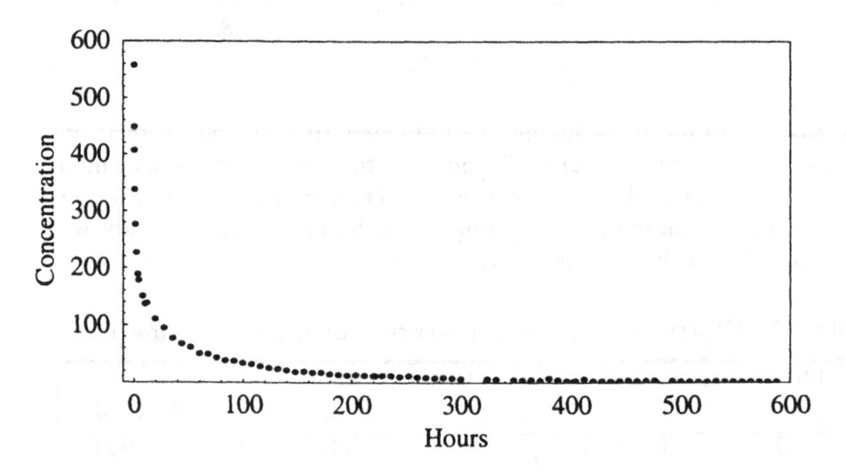

Table 2.4. Concentration of labelled calcium in plasma of an adult woman.

Time (hr)	Concentration	Time (hr)	Concentration
0.082	557.54	243.5	10.16
0.175	448.18	251.5	11.50
0.25	406.89	259.5	9.45
0.5	337.53	267.5	8.98
1.0	276.53	275.5	7.92
2.0	227.06	283.5	8.71
3.5	188.89	291.5	7.71
4.083	178.62	299.5	6.55
8.0	151.04	323.5	6.34
10.0	137.32	331.5	5.78
11.5	138.44	347.5	4.79
19.5	110.78	355.5	4.39
27.5	95.05	363.5	5.00
35.5	77.07	371.5	3.88
43.5	67.41	379.5	7.54
51.5	61.46	387.5	4.85
59.5	50.27	395.5	3.05
67.5	49.31	403.5	3.24
75.5	43.30	411.5	6.10
83.5	38.34	419.5	2.82
91.5	37.21	427.5	2.70
99.5	33.81	435.5	2.48
107.5	31.53	443.5	2.96
115.5	27.57	451.5	3.73
123.5	25.58	459.5	3.33
131.5	23.64	467.5	3.35
139.5	21.21	475.5	3.97
147.5	18.50	491.5	3.11
155.5	19.02	499.5	2.04
163.5	17.27	507.5	2.16
171.5	17.34	515.5	1.37
179.5	14.64	523.5	1.68
187.5	13.43	531.5	2.97
195.5	12.53	539.5	1.86
203.5	13.25	547.5	1.92
211.5	12.26	555.5	1.97
219.5	11.85	563.5	1.97
221.0	11.71	571.5	1.58
227.5	12.06	579.5	1.75
235.5	11.81	587.5	1.67

Part II Models for a Single Population

3

Basic Methodology for Single Population Stochastic Models

3.1 Introduction

This chapter introduces the basic theoretical tools utilized subsequently for solving single population stochastic models. Most of these tools are developed more extensively in the basic textbooks in stochastic processes, including the leading applied texts by Bailey [4], Chiang [12] and Renshaw [82]. The basic methodology is also described and illustrated specifically as applied to standard compartmental models in Jacquez [28]. This monograph will build on Jacquez' development, most notably by including "births" into the stochastic compartmental models.

The basic notation and assumptions are introduced in Section 3.2. Section 3.3 discusses the use of cumulants which will be utilized extensively. Section 3.4 describes the solution of stochastic models using a system of so-called Kolmogorov equations. Section 3.5 presents some generating functions of interest, which are then utilized in Section 3.6 to solve the stochastic models through the use of pertinent partial differential equations. Concluding remarks concerning these theoretical tools follow in Section 3.7.

3.2 Basic Assumptions

Let

$X(t)$ denote the random population size at elapsed time t, with

$p_x(t) = \text{Prob}[X(t) = x]$, i.e. the probability that the population size is x at time t, and

$\mathbf{p}(t) = [p_0(t), p_1(t), \ldots, p_x(t), \ldots]$, the probability distribution of $X(t)$.

$$(3.1)$$

One general objective is to solve for $\mathbf{p}(t)$, for any $t > 0$, from simple assumptions concerning the dynamics of the population.

In the standard single population birth-death (BD) models, there are three possible types of changes over time, namely immigration, birth and death, in the population size, $X(t)$. The "instantaneous" probabilities of these possible *unit* changes in small intervals of time from t to $t + \Delta t$ are assumed to be *independent* with:

$$\begin{aligned}
&1.\ \text{Prob}\,\{X \text{ will increase by 1 due to immigration}\} = I\Delta t,\\
&2.\ \text{Prob}\,\{X \text{ will increase by 1 due to birth}\} = \lambda_X \Delta t, \text{ and}\\
&3.\ \text{Prob}\,\{X \text{ will decrease by 1 due to death}\} = \mu_X \Delta t.
\end{aligned} \tag{3.2}$$

We will consider a very special case of the assumptions in (3.2) in order to illustrate the basic methodology for analyzing single population models. The special case is the so-called linear immigration-death model, which is equivalent to a simple stochastic one-compartment model. The model is considered in many texts on stochastic processes, and is also developed in the compartmental modeling literature by, among others, Jacquez [28]. The subsequent chapters will develop models with other assumed forms of the rates in (3.2), using primarily the methodology outlined in this chapter.

The immigration rate in the model of present interest is assumed to be some constant, denoted I, and the death rate is assumed to be proportional to population size, i.e.

$$\mu_X = aX. \tag{3.3}$$

For simplicity, the initial population size is assumed to be

$$X(0) = 0. \tag{3.4}$$

The assumption of zero initial size in (3.4) will be relaxed in Chapter 4 in order to enhance the versatility of the model. However even the present simple model is useful in some applications. As an illustration, let $X(t)$ denote the number of insects of a given species in a field at time t. For example, the insect could be the corn earworm which migrates in the spring, transported by wind, through the central U.S. [101]. Suppose a field has no earworm at time 0, i.e. $X(0) = 0$, and that they immigrate to the field at rate $I = 10$ insects/day. Suppose, moreover, that the departure (or "death") rate of the insect population is $\mu_X = 0.1\,X$. This will subsequently be interpreted as an exponential "waiting-time" distribution, with a mean stay of $(0.1)^{-1} = 10$ days for any given insect in the field.

For subsequent comparative purposes, consider first the deterministic formulation of this model. Letting $\dot{X}\ (t)$ denote the derivative of size $X(t)$, the deterministic model is

$$\dot{X}\ (t) = I - aX, \tag{3.5}$$

with solution

$$X(t) = (1 - e^{-at})I/a. \tag{3.6}$$

For the specific corn earworm population illustration, the solution for population size is

$$X(t) = 100(1 - e^{-0.1t}). \tag{3.7}$$

Analogous stochastic solutions follow in the subsequent sections.

3.3 Moments and Cumulants

In a stochastic model, the population size $X(t)$ is a random variable for any $t > 0$, hence, for any t, $X(t)$ would have a probability density function, denoted $\mathbf{p}(t)$ in (3.2). We consider first some convenient summary measures for probability density functions.

The statistical "moments" are widely used numerical summary measures of probability distributions. Specifically let

$$\mu_i(t) = \Sigma x^i p_x(t) \tag{3.8}$$

denote the ith moment (about the origin) of $X(t)$. In the present applications to non-negative (counting) distributions, the moments usually become very large numerically for $i > 1$. One solution to this problem is to use "central moments", i.e. moments of the variable $X(t) - \mu_1(t)$.

Another approach, which is used extensively in this monograph, is to describe the distributions through the use of so-called "cumulants". Let $\kappa_i(t)$ denote the ith cumulant of $X(t)$. A "cumulant" will be formally defined in Section 3.6. For most practical purposes, it suffices to note that the first three cumulants of $X(t)$ are the mean, variance and skewness, respectively, for $X(t)$.

The general relationship between cumulants and moments is given in [30, 90] as:

$$\mu_m = \sum_{j=0}^{m-1} \binom{m-1}{j} \kappa_{m-j}\mu_j \tag{3.9}$$

with $\mu_0 \equiv 1$. In particular, the first three cumulant functions may be obtained from (3.9) as

$$\begin{aligned} \kappa_1(t) &= \mu_1(t) \\ \kappa_2(t) &= \mu_2(t) - \mu_1^2(t) \\ \kappa_3(t) &= \mu_3(t) - 3\mu_1(t)\mu_2(t) + 2\mu_1^3(t) \end{aligned} \tag{3.10}$$

which, as noted, give the mean, variance and skewness functions for $X(t)$ from the $\mu_i(t)$ moment functions.

There are a number of advantages to using cumulants instead of moments, even though the former are not as well-known. One is that the first three cumulants have obvious appeal as a convenient three-number summary of a distribution. As Whittle [103] states: "We prefer to deal with cumulants rather than the moments, since these give a more obvious characterization of the variate." A corollary to this is that the cumulant structure provides a convenient characterization for some common distributions. Two outstanding cases of this are: 1) for the Poisson distribution, all cumulants are equal, i.e. $\kappa_i = c$ for all i, and 2) for the Normal distribution, all cumulants above order two are zero, i.e. $\kappa_i = 0$ for $i > 2$.

A second advantage is that cumulant functions provide a basis for parameter estimation using least squares. The mean value function, $\kappa_1(t)$, could serve as the regression model; the variance function, $\kappa_2(t)$, gives the weights; and the skewness function, $\kappa_3(t)$, provides a simple indicator of possible departure from an assumed Normal (hence symmetric) distribution.

A third advantage is that low-order cumulants may be utilized to give saddlepoint approximations of the underlying distribution. Renshaw [83] develops the use of saddlepoint approximations specifically from the cumulants of stochastic processes. In particular, he gives the following completely general and tractable approximation based on using exact cumulants up to third-order (i.e. $\kappa_1 = \mu$, $\kappa_2 = \sigma^2$ and κ_3):

$$f(x) = (4\pi^2\psi)^{-1/4} \exp\left\{-(1/6\kappa_3^2)\left[\sigma^3 - 3\sigma^2\psi + 2\psi^{3/2}\right]\right\} \quad (3.11)$$

where $\psi = \sigma^4 + 2\kappa_3(x - \mu)$. His work shows, in principle, that adequate approximations of size distributions may be constructed from low order exact cumulants, indeed the approximation in (3.11) is shown to approximate a Poisson distribution well. Approximations based on higher-order cumulants are also given.

In summary, we have found cumulant functions to be more convenient in our practical model building than the corresponding moment functions. As an illustration of using cumulants, consider the simple immigration-death model proposed in Section 3.2. It will be shown later that its population size, $X(t)$, has a Poisson distribution with parameter function

$$\lambda(t) = (1 - e^{-at})I/a. \quad (3.12)$$

Therefore, the probability function for any $t > 0$ is given, by definition, as

$$p_x(t) = \exp\{-\lambda(t)\}[\lambda(t)]^x/x! \quad (3.13)$$

for $x \geq 0$.

The moments for $X(t)$ may be derived directly by substituting (3.13) into (3.8). However, the low order moments may also be obtained by using simple properties of the Poisson distribution which give the moments as a function of the Poisson parameter. Using these relationships, in e.g. [30], one can show that the moments are

$$\mu_1(t) = \lambda(t)$$
$$\mu_2(t) = \lambda(t)[1 + \lambda(t)], \text{ and}$$
$$\mu_3(t) = \lambda(t)\{[\lambda(t) + 1]^2 + \lambda(t)\}.$$

The cumulant functions follow from (3.10) as

$$\kappa_1(t) = \kappa_2(t) = \kappa_3(t) = \lambda(t) = (1 - e^{-at})I/a, \quad (3.14)$$

which illustrates the useful characterization of the Poisson distribution noted previously, namely the equality of all of its cumulants.

The parameter function for the corn earworm migration problem with $I = 10$ and $a = 0.1$ is clearly, from (3.12):

$$\lambda(t) = 100(1 - e^{-0.1t}),$$

which is the deterministic solution in (3.7). The mean value, variance and skewness functions from (3.14) are immediately

$$\kappa_1(t) = \kappa_2(t) = \kappa_3(t) = 100(1 - e^{-0.1t}).$$

The probabilities of various population sizes, $X(t)$, for given t are easily found by evaluating $\lambda(t)$ at the desired t and substituting into (3.13).

As a special case, consider the equilibrium size distribution which is Poisson (100). Clearly, the mean, variance and skewness for this equilibrium distribution are all equal to 100. Supppose that we had derived these cumulant values, but did not know the exact distribution, as is often the case in the subsequent, more involved models. A simple saddlepoint approximation could be obtained by substitution into (3.11). These results are illustrated in Figure 3.1. The accuracy of the approximation is not surprising in light of the near-Normality of this Poisson distribution.

3.4 Kolmogorov Differential Equations

One standard approach for solving for the probability functions, such as (3.13), generated by various stochastic processes is to use equations known as the Kolmogorov differential equations. This method will be illustrated only for the present immigration-death model. The Kolmogorov equations for this model are developed in [28, p. 239-241], and the example is sufficient for our purposes to illustrate all major points of the method. The latter chapters will use this methodology extensively to analyze subsequent models.

Consider the event of the population size being x at time $t + \Delta t$, where Δt is some small time interval. In symbols, this event is $X(t + \Delta t) = x$. There are a number of mutually exclusive ways in which this event could have come about, starting from time t. Specifically, they are: i) to have size x at time t with no change from t to $t + \Delta t$, ii) to have size $x + 1$ at time t with only a single death in the next interval Δt, iii) to have size $x - 1$ at time t with only a single immigration in Δt, and iv) other ways which involve two or more independent changes of unit size in the interval Δt. Because this set of mutually exclusive "pathways" to the desired event at $t + \Delta t$ is exhaustive, the probability of size $x + 1$ at $t + \Delta t$ may be written as the sum of the

Figure 3.1. Comparison of the (discrete) Poisson (100) distribution with its (continuous) saddlepoint approximation.

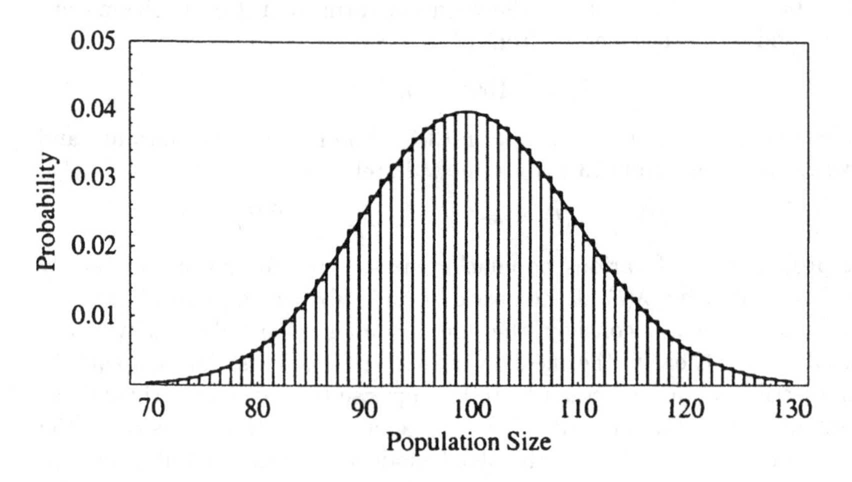

individual probabilities of these pathways. Symbolically, using the assumptions for possible changes with constant immigration I and deaths in (3.3), one has for suitably small Δt:

$$p_x(t+\Delta t) = p_x(t)[1 - I\Delta t - ax\Delta t] + p_{x+1}(t)[a(x+1)\Delta t] \\ + p_{x-1}(t)[I\Delta t] + o(\Delta t), \qquad \text{for } x > 0,$$

where $o(\Delta t)$ denotes terms of higher order Δt associated with multiple independent changes. Subtracting $p_x(t)$, dividing by Δt, and taking the limit as $\Delta t \to 0$, one has

$$\dot{p}_x\,(t) = Ip_{x-1}(t) - (I + ax)p_x(t) + a(x+1)p_{x+1}(t) \text{ for } x > 0, \quad (3.15)$$

with

$$\dot{p}_0\,(t) = -Ip_0(t) + ap_1(t). \quad (3.16)$$

The solution of this set of differential equations yields the desired (transient) probability distribution in (3.1) for $X(t)$.

As an example, consider our model with initial condition $X(0) = 0$, which implies $p_0(0) = 1$ with $p_i(0) = 0$ for $i > 0$. One could verify that the solution given in (3.13) satisfies the Kolmogorov equations in (3.15) and (3.16). This implies that $X(t)$ has a Poisson distribution with parameter function $\lambda(t)$ in (3.11), symbolically

$$X(t) \sim \text{ Poisson } \{(1 - e^{-at})I/a\}.$$

In practice, analytical solutions to Kolmogorov equations such as (3.15) and (3.16) are seldom obvious. Other methods for obtaining the analytical solutions follow later. For cases where the analytical solution is not available, note that one could consider finding the direct numerical solution of equations (3.15) and (3.16) for particular parameter values of interest.

It is also helpful for subsequent considerations to express the Kolmogorov equations in matrix form. Recall from (3.1) that $\mathbf{p}(t)$ is the row vector of $p_x(t)$ for $x = 0, 1, \ldots$. The Kolmogorov equations for a single population model with the standard assumptions for unit changes given in (3.2) may be written in the form

$$\dot{\mathbf{p}}\ (t) = \mathbf{p}(t)\mathbf{R}, \tag{3.17}$$

where $\mathbf{R}$ is a tridiagonal matrix. For example, the $\mathbf{R}$ matrix for the system of equations in (3.15) and (3.16) is infinite, with elements (for $i,\ j \geq 0$):

$$r_{i,j} = \begin{cases} r_{i,i+1} = I \\ r_{i,i-1} = ai \\ r_{i,i} = -(ai + I) \\ r_{i,j} = 0 & \text{for } |i - j| > 1. \end{cases} \tag{3.18}$$

As a specific illustration, the $\mathbf{R}$ matrix for the corn earworm problem is:

$$\mathbf{R} = \begin{bmatrix} -10 & 10 & 0 & \ldots \\ 0.1 & -10.1 & 10 & \ldots \\ 0 & 0.2 & -10.2 & \ldots \\ 0 & 0 & 0.3 & \ldots \\ \vdots & \vdots & \vdots & \end{bmatrix}$$

For models with finite $\mathbf{R}$ in (3.17), one can proceed to find numerical solutions for the transient probabililty distributions by the direct solution of the differential equations. In the present immigration-death problem, however, the infinite $\mathbf{R}$ matrix rules out a direct "exact" solution. One option for such problems is to truncate the set of differential equations for some large upper bound on population size, and then proceed to find directly a close approximate solution for the size distributions. This useful option is illustrated in subsequent chapters. Another option is to obtain the exact solution by using methodology described in the following two sections.

3.5 Generating Functions

Generating functions are coming into widespread usage as methodological tools. They may be used to obtain the population size distribution in analytical form for simple population models, and also to solve directly for the cumulant functions of more involved models. The authors support the assertion that "the beginner needs to know their definitions and the more advanced student should learn how to use them" [28, p. 245]. Generating functions are used extensively throughout this monograph.

The random variable, $X(t)$, for population size is non-negative integer-valued. The probability generating function, or pgf, denoted $P(s,t)$, for such a random variable may be defined as

$$P(s,t) = \Sigma s^x p_x(t), \tag{3.19}$$

where s is a "dummy variable" such that $|s| < 1$. It follows that one could obtain any probability, say $p_i(t)$, by differentiating $P(s,t)$, specifically

$$p_i(t) = d^i P(s,t)/ds^i|_{s=0}$$

or in other notation as

$$p_i(t) = P^{(i)}(0,t), \tag{3.20}$$

where $P^{(i)}(0,t)$ denotes the ith derivative evaluated at $s = 0$.

The moment generating function, or mgf, denoted $M(\theta,t)$, may be defined for an integer-valued variable as

$$M(\theta,t) = \Sigma e^{\theta x} p_x(t), \tag{3.21}$$

where θ is a dummy variable. Clearly, using (3.19), one has

$$M(\theta,t) = P(e^\theta, t). \tag{3.22}$$

It can be shown, by expanding $e^{\theta x}$ in (3.21) and using (3.8), that $M(\theta,t)$ may be expressed as the power series

$$M(\theta,t) = \sum_{i\geq 0} \mu_i(t)\theta^i/i! \tag{3.23}$$

with $\mu_0 = 1$. It follows that the ith moment may be obtained from the mgf, similar to (3.20), as

$$\mu_i(t) = M^{(i)}(0,t). \tag{3.24}$$

The cumulant generating function, or cgf, denoted $K(\theta,t)$ is defined as

$$K(\theta,t) = \log M(\theta,t), \tag{3.25}$$

with power series expansion

$$K(\theta,t) = \sum_{i\geq 1} \kappa_i(t)\theta^i/i!. \tag{3.26}$$

Equation (3.26) formally *defines* a cumulant, $\kappa_i(t)$, as a coefficient in the series expansion of $K(\theta,t)$. It too is easily found from its generating function as

$$\kappa_i(t) = K^{(i)}(0,t). \tag{3.27}$$

In our simple illustration, the probability generating function for the Poisson population size variable in (3.12) for the immigration-death model may be shown using (3.19) to be:

$$P(s,t) = \exp\{(s-1)\lambda(t)\}, \text{ where} \qquad (3.28)$$
$$\lambda(t) = (1 - e^{-at})I/a.$$

Using (3.22), the moment generating function is:

$$M(\theta,t) = \exp\{(e^{\theta}-1)\lambda(t)\}.$$

The moments (3.13) could be obtained from this mgf using the repeated differentiation in (3.24). The cumulant generating function is easily found using definition (3.25) as

$$K(\theta,t) = (e^{\theta}-1)\lambda(t).$$

The cumulants, using (3.27), follow as:

$$\kappa_i(t) = \lambda(t) \quad \text{for all } i,$$

i.e. that *all* cumulants are identical for the Poisson distribution, as illustrated in (3.14).

As an example, consider the corn earworm population modeling problem. The equilibrium size distribution, for $X(\infty)$, has parameter value $\lambda = 100$. The generating functions for this case follow from (3.28), (3.22) and (3.25) as

$$P(s) = \exp\{100(s-1)\}$$
$$M(\theta) = \exp\{100(e^{\theta}-1)\}$$
$$K(\theta) = 100(e^{\theta}-1)$$

The reason for using generating functions is not yet transparent. In subsequent work, however, methods exist whereby one could find a generating function directly from model assumptions, without first finding the probabilities. Indeed, one may show using generating functions that the cumulant functions for many models are relatively simple even though the corresponding probability functions are intractable.

3.6 Partial Differential Equations for Generating Functions

A very useful tool for finding analytically the distribution of $X(t)$ and/or its cumulants is to obtain and solve partial differential equations for the associated generating functions. This is illustrated for the present immigration-death model in [28, p. 245]. Consider multiplying equations (3.15) and (3.16) by s^x and summing over x, to give

$$\sum s^x \dot{p}_x = I\sum s^x p_{x-1} - \sum (I+ax)s^x p_x + a\sum (x+1)s^x p_{x+1}.$$

The left-hand side, using (3.19), is $\partial P(s,t)/\partial t$. Carrying out the summations on the right, using the additional result

$$\partial P(s,t)/\partial s = x \sum s^{x-1} p_x(t)$$

also obtained from (3.19), yields after some manipulation

$$\frac{\partial P(s,t)}{\partial t} = I(s-1)P(s,t) + a(1-s)\frac{\partial P(s,t)}{\partial s}. \tag{3.29}$$

The initial condition corresponding to $X(0) = 0$ is obtained from (3.19) as $P(s,0) = 1$. The solution to this linear partial differential equation is

$$P(s,t) = \exp\left\{(s-1)(1-e^{-at})I/a\right\},$$

as previously given in (3.28) and in [28, p. 245]. This result proves that $X(t)$ has a Poisson distribution, as claimed in (3.13).

Partial differential equations corresponding to (3.29) may be written directly for birth-death-migration models using an infinitesimal generator technique, called the "random-variable" technique, given in [4]. Let the possible changes in population size $X(t)$ from t to $t + \Delta t$ be denoted as

$$\text{Prob}\{X(t) \text{ changes by } j \text{ units}\} = f_j(X)\Delta t + o(\Delta t) \tag{3.30}$$

For example, in terms of these f_j "intensity functions," the possible changes for the immigration-death model are

$$f_1 = I \quad \text{and} \quad f_{-1} = ax \tag{3.31}$$

where the first function implies that the rate of unit additions to the population is the constant I, and the second function implies that the rate of unit departures is a linear function of X, as assumed in (3.3).

For intensity functions of form

$$f(x) = \sum a_k X^k,$$

we define the operator notation

$$f\left(s\frac{\partial}{\partial s}\right)P = \sum a_k s^k \frac{\partial^k P}{\partial s^k} \quad \text{and}$$

$$f\left(\frac{\partial}{\partial \theta}\right)M = \sum a_k \frac{\partial^k M}{\partial \theta^k}.$$

Using this notation, the following operator equations for the pgf and mgf are given in [4, p. 73]:

$$\frac{\partial P}{\partial t} = \sum_{j\neq 0}(s^j - 1)f_j\left(s\frac{\partial}{\partial s}\right)P(s,t) \tag{3.32}$$

$$\frac{\partial M}{\partial t} = \sum_{j\neq 0}(e^{j\theta} - 1)f_j\left(\frac{\partial}{\partial \theta}\right)M(\theta,t). \tag{3.33}$$

As an illustration substituting the terms in (3.31), of degrees 0 and 1 with respect to X respectively, into (3.32) and (3.33) yields

$$\frac{\partial P}{\partial t} = I(s-1)P(s,t) + a(1-s)\frac{\partial P}{\partial s}, \tag{3.34}$$

as in (3.29), and

$$\frac{\partial M}{\partial t} = I(e^{\theta}-1)M(\theta,t) + a(e^{-\theta}-1)\frac{\partial M}{\partial \theta}. \tag{3.35}$$

Note that each of the two possible changes in $X(t)$ leads to a term on the right in (3.34) and (3.35). One could verify that the solution to (3.35), with boundary condition $M(\theta,0) = 1$ from (3.21), is:

$$M(\theta,t) = \exp\{(e^{\theta}-1)(1-e^{-at})I/a\} \tag{3.36}$$

as given previously.

The use of the operator equations in (3.32) and (3.33) is a very useful approach. The approach may be applied easily to density-dependent models, for which the intensity functions involve higher powers of X leading to nonlinear partial differential equations, as demonstrated in Chapters 6 and 7. The approach also extends to multiple populations, which will be illustrated in Part III.

Equations (3.34) and (3.35), which characterize this immigration-death model, are linear with the relatively easy solutions in (3.28) and (3.36). In cases where such analytical solutions are not available, one might consider using series expansions of (3.34) and (3.35). For example, substituting the series expansion (3.19) into (3.34), one would obtain differential equations for the probabilities, i.e. the Kolmogorov equations in (3.15) and (3.16). Similarly, substituting (3.21) into (3.35) would give differential equations for the moments.

To illustrate this process, consider finding differential equations for the cumulants. Substituting definition (3.25), i.e.

$$K(\theta,t) = \log M(\theta,t)$$

into (3.35), one has the following equation for the cgf:

$$\frac{\partial K}{\partial t} = I(e^{\theta}-1) + a(e^{-\theta}-1)\frac{\partial K}{\partial \theta}. \tag{3.37}$$

Using the series expansion (3.26), and equating coefficients of powers of θ it follows that:

$$\begin{aligned}
\dot{\kappa}_1(t) &= I - a\kappa_1(t) \\
\dot{\kappa}_2(t) &= I + a\kappa_1(t) - 2a\kappa_2(t) \\
\dot{\kappa}_3(t) &= I - a\kappa_1(t) + 3a\kappa_2(t) - 3a\kappa_3(t).
\end{aligned} \tag{3.38}$$

With the initial condition $\kappa_1(0) = \kappa_2(0) = \kappa_3(0) = 0$, corresponding to $X(0) = 0$, the solution to these cumulant equations is

$$\kappa_1(t) = \kappa_2(t) = \kappa_3(t) = (1 - e^{-at})I/a, \tag{3.39}$$

as given in (3.14).

Note how simple this approach to finding differential equations for cumulants is to use in practice. The model assumptions in (3.31) were substituted directly into operator equation (3.33), which was transformed by (3.25) to yield (3.37). The expansion of (3.37) gives (3.38). For this model, the solutions in (3.39) are simple. The same procedure will be followed subsequently with other models to obtain analogous differential equations, which will be solved numerically if analytical solutions are not tractable.

3.7 General Approach to Single Population Growth Models

This chapter used a simple immigration-death model to illustrate the theoretical development of the model. Consider now a much broader framework in which population birth and death rate functions in (3.2) are defined as polynomials of the form:

$$\lambda_X = a_1 X - b_1 X^{s+1} \qquad \text{and} \qquad \mu_X = a_2 X + b_2 X^{s+1}, \tag{3.40}$$

respectively, where a_1, a_2, b_1, b_2, and $s > 0$. The "per capita" rates, also called the specific rates, are immediately

$$\lambda_X / X = a_1 - b_1 X^s$$
$$\mu_X / X = a_2 + b_2 X^s$$

In these equations, the a_i are interpreted as the "intrinsic rates" and the b_i as the "crowding coefficients." Chapters 4 and 5 discuss the case of $b_i \equiv 0$ for $i = 1, 2$ in (3.40). Therefore, their rate functions are (simple) linear and the corresponding population models are called density-independent. Chapters 6 and 7 develop the case of $b_i > 0$ for some i, with $s \geq 1$. The resulting rate functions are nonlinear and the models are called density-dependent.

The methodology outlined in this chapter will be used to analyze these various single population models as follows. The model of interest will first be introduced, and its deterministic solution given. The Kolmogorov equations for the stochastic model similar to Section 3.4 will follow, and their numerical solutions will be obtained and explored, where possible. The partial differential equations for the cumulant generating functions will then be analyzed, and its cumulant functions obtained and investigated.

After this theoretical development, the models are applied to various examples, particularly to the AHB and the muskrat spread examples, where appropriate. Finally, most chapters will discuss a number of extensions, including methodology unique to the particular model, which either simplifies the computations or contributes new insight into the underlying kinetic structure of the problem.

4

Linear Immigration-Death Models

4.1 Introduction

Consider first modeling a single population of size $X(t)$, with the linear death rate

$$\mu_X = aX \tag{4.1}$$

and immigration rate I. This very simple case of the general rates in (3.40) extends the illustration in Chapter 3 by relaxing the assumption of an initial population size zero. The model, as previously noted, is called the linear immigration-death (LID) model, and it is widely used as the standard stochastic one-compartmental model. This model is developed thoroughly in [28] in a compartmental context. A simple application developed subsequently is modeling the bioaccumulation of mercury, described in Section 2.4. We have also used the model extensively to describe the passage of digesta in ruminant animals, see e.g. [15, 16, 59].

An application of an entirely different nature is population size modeling, where the $X(t)$ size variable is conceptualized as an actual discrete, counting variable. In this context, the LID model is useful as a basic model for the size of a population without a birth mechanism. This chapter has such application to AHB dynamics, as illustrated subsequently.

4.2 Deterministic Model

4.2.1 Solution to Deterministic Model

The deterministic model is simply

$$\dot{X}(t) = -aX + I. \tag{4.2}$$

Its solution, with initial value $X(0) = X_0$, is

$$X(t) = X_0 e^{-at} + (1 - e^{-at}) I/a. \tag{4.3}$$

In common practice, model (4.3) is fitted to data to estimate the unknown parameters, which might include X_0, I and a.

Let X^* denote the equilibrium size, i.e. $X^* = \lim_{t\to\infty} X(t)$. It is easy to show from either (4.2) or (4.3) that

$$X^* = I/a \tag{4.4}$$

4.2.2 Application to Bioaccumulation of Mercury in Fish

Consider applying this deterministic model to the mercury accumulation data. The fish were initially free of mercury i.e. $X(0) = 0$. The regression function, obtained by dividing the deterministic model (4.3) by V, the fish size, is

$$c(t) = I^*(1 - e^{-at})/a \tag{4.5}$$

where I^* is the standardized uptake (immigration) rate I/V.

The fitted curves are illustrated in Figure 4.1. The least squares estimates of the mercury elimination rate, a, for the three fish are 0.62, 0.39 and $0.62d^{-1}$. Under this simple LID model which assumes that the fish is a "homogeneous" compartment, the rate for the whole group of 21 fish varies from roughly 39% to 83% of the fish's mercury accumulation per day. The estimates of I^* for the three fish are 0.31, 0.32 and 0.60, μg Hg/gm dry wt/day with the mercury uptake for the group of 21 fish ranging from 0.11 to 0.60. The estimated equilibrium concentrations, $c(\infty) = I^*/a$, are 0.50, 0.83 and 0.97 μg Hg/gm dry wt. for the three fish, with a range of 0.22 to 1.00 for the group of 21.

Figure 4.1. Observed and fitted values for the LID model of mercury bioaccumulation over time in three individual fish.

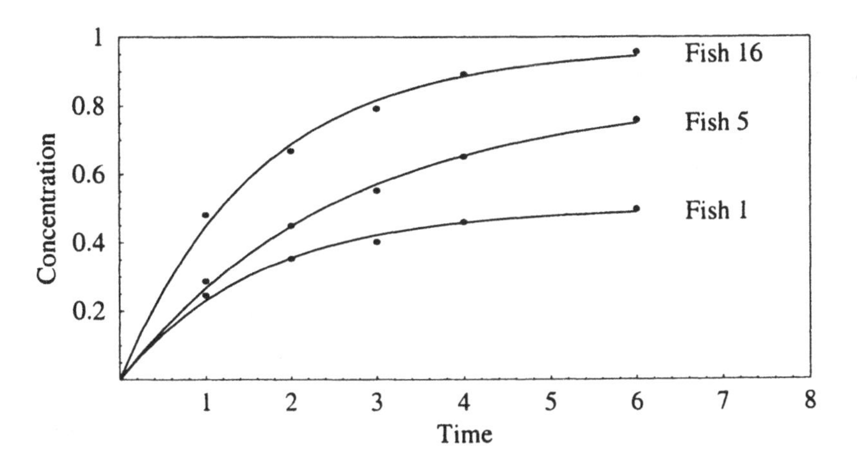

Though the model fits all of the fish data fairly well, with details given in [57], the residuals for any individual fish tend not to be randomly scattered. An obvious extension would be to fit more elaborate, multicompartment models developed in Chapter 10 to the data, in order to provide a mercury "storage" compartment.

4.3 Probability Distributions for the Stochastic Model

4.3.1 Kolmogorov Equations

The derivation of the Kolmogorov forward equations was previously given in Section 3.4 for the LID model. In matrix notation, this system of equations was given as

$$\dot{\mathbf{p}}\ (t) = \mathbf{p}(t)\mathbf{R}, \tag{4.6}$$

where $\mathbf{R}$ is a tridiagonal matrix with elements

$$r_{i,j} = \begin{cases} r_{i,i+1} = I \\ r_{i,i-1} = ai \\ r_{i,i} = -(ai + I) \\ r_{i,j} = 0 & \text{otherwise.} \end{cases}$$

Although this is an infinite system of equations, the probabilities of sizes much larger than max $[X_0, I/a]$ become minute, as noted in [28, p. 241]. Therefore one can truncate the distribution of $X(t)$ at some large upper value, say U, and solve the resulting finite system in (4.6) for numerical approximations of the $p_i(t)$ probability functions. These approximations are usually extremely accurate for large U.

4.3.2 Equilibrium Distribution

The limiting distribution for the equilibrium population size, X^*, is of great interest in modeling, and for this model the equilibrium distribution may be obtained directly from the Kolmogorov equations. Assuming that the distribution exists, we set $\dot{\mathbf{p}}\ (t) = \mathbf{0}$ in (4.6) or equivalently in (3.15) and (3.16). Letting π_i denote $p_i(\infty)$, it follows from (3.16) that

$$\pi_1 = \pi_0(I/a).$$

Substitution into (3.15) yields

$$\pi_2 = \pi_0(I/a)^2/2,$$

and in general

$$\pi_i = \pi_0(I/a)^i/i! \tag{4.7}$$

Summing π_i over all i, and adding π_0 to both sides, one has

$$1 = \pi_0 e^{I/a},$$

whereupon

$$\pi_0 = e^{-I/a}.$$

Using (4.7), it follows that

$$\pi_i = e^{-I/a}(I/a)^i/i! \tag{4.8}$$

Symbolically,

$$X^* \sim \text{ Poisson } (I/a). \tag{4.9}$$

4.4 Generating Functions

The partial differential equation which characterizes the stochastic model through the probability generating function (pgf) was obtained both directly from the Kolmogorov equations in (3.29) and through the operator equations in (3.34). This equation,

$$\frac{\partial P}{\partial t} = a(1-s)\frac{\partial P}{\partial s} + I(s-1)P \tag{4.10}$$

was solved assuming $X(0) = 0$ in Chapter 3. More generally, the solution for $X(0) = X_0$ is

$$P(s,t) = [1 + (s-1)e^{-at}]^{X_0} \exp\left\{(s-1)(1-e^{-at})I/a\right\}. \tag{4.11}$$

The analytical solution is useful for the following reasons:

i) Result (4.11) may be used to derive expressions for the individual $p_i(t)$ probability functions. Consider first the limiting distribution of X^*. Because $a > 0$, it follows from (4.11) that the generating function $P(s, \infty)$ of X^* is clearly:

$$P(s, \infty) = \exp\left\{(s-1)I/a\right\}.$$

which may be written in series form as

$$P(s, \infty) = e^{-I/a}\sum_i (sI/a)^i/i!$$

This implies from (3.19) that X^* has the Poisson distribution with parameter $\lambda = I/a$, as noted in (4.8) and (4.9). Clearly, this limiting distribution is independent of initial size, X_0.

Note that (4.11) has two factors. The first factor is the generating function of a binomial distribution with parameters $n = X_0$ and $p = e^{-at}$ [30]. This distribution represents the number of the X_0 particles which are still in the compartment at time t. Let this random variable be denoted $X^{(1)}(t)$. The second factor is the generating function of a Poisson distribution with parameter $\lambda = (1 - e^{-at})I/a$, as given in (3.28). This distribution represents the number of particles which arrived after time 0 and are still in the compartment. Let this random variable be denoted $X^{(0)}(t)$. Because the generating function of $X(t)$ in (4.11) is the product of these individual generating functions, it follows from statistical theory that $X(t)$ is the sum of the two independent variables [74]. Symbolically

$$\begin{aligned} X(t) &= X^{(0)}(t) + X^{(1)}(t) \quad \text{where} \\ X^{(1)}(t) &\sim \text{binomial } [n = X_0, p = e^{-at}] \text{ and} \\ X^{(0)}(t) &\sim \text{Poisson } [\lambda = (1 - e^{-at})I/a]. \end{aligned} \tag{4.12}$$

This structure will be useful subsequently. Analytical expressions for the $p_i(t)$, which define the probability distribution of $X(t)$ at any t, could also be obtained, but they are not of present interest.

ii) Though the $p_i(t)$ expressions are unwieldy in general, the probability that the population size is 0, i.e. $X(t) = 0$, is by itself very useful and illuminating. It is easily found by setting $i = 0$ in (3.20), whereupon

$$p_0(t) = P(0, t). \tag{4.13}$$

For this LID model, it follows immediately from (4.11) that

$$p_0(t) = (1 - e^{-at})^{X_0} \exp\left\{(e^{-at} - 1)I/a\right\}. \tag{4.14}$$

Its equilibrium value is

$$p_0(\infty) = e^{-I/a},$$

which also follows directly from (4.8). Result (4.14) will be illustrated subsequently.

iii) The moment generating function, $M(\theta, t)$, may be obtained for the general case $X(0) = X_0$ by substituting e^θ for s in (4.11). The cumulant generating function $K(\theta, t) = \log M(\theta, t)$ follows as

$$K(\theta, t) = (e^\theta - 1)(1 - e^{-at})I/a + X_0 \log\left[1 + (e^\theta - 1)e^{-at}\right]. \tag{4.15}$$

The series expansion of $K(\theta, t)$ gives the cumulant functions, which will be illustrated in the next section.

Clearly, the analytical solution for the pgf is useful when it is available, as in (4.11) for the present model. If the primary objective had been to obtain the cgf in (4.15), one could have proceeded to write first the partial differential equation for the moment generating function, as illustrated in (3.35) in Chapter 3. The transformation to the cumulant generating function yields

$$\frac{\partial K}{\partial t} = a(e^{-\theta} - 1)\frac{\partial K}{\partial \theta} + I(e^\theta - 1), \tag{4.16}$$

previously given in (3.37). This equation, with initial condition $K(\theta, 0) = X_0$, could be solved directly to give the analytical solution in (4.15).

In subsequent models, the analytical solutions to the pde corresponding to (4.10) and (4.16) for the pgf and cgf, respectively, either do not exist or are far too unwieldy to be useful in practice. However, as illustrated in Section 3.6, one could substitute the series expansion of the generating function, i.e. either (3.19) or (3.26), into the appropriate pde to obtain directly differential equations for the desired probabilities or cumulants. For example, substituting (3.19) into (4.10), and equating coefficients of powers of s on both sides of the equation yields the Kolmogorov equations in (4.6). Differential equations for the cumulant functions could be obtained in similar fashion from (4.16), as indicated in the next section.

4.5 Cumulant Functions

In general, the analytical solutions for the transient probability distributions, $\mathbf{p}(t)$, are unwieldy, though numerical solutions are available, of course. Consider now the use of cumulant functions. They are often available in analytical form, with surprisingly simple solutions, and they often provide great insight into the stochastic solutions for the various assumed population models.

Cumulant functions for the LID model may be obtained through any of the following three general approaches:

i) Using basic relationships for the cumulants of some well-known distributions, in this case for binomial and Poisson distributions [30], it follows directly from (4.12) that the first three cumulant functions for $X(t)$ are:

$$\mu(t) = np + \lambda = X_0 e^{-at} + (1 - e^{-at}) I/a \tag{4.17}$$

$$\sigma^2(t) = np(1-p) + \lambda = \mu(t) - X_0 e^{-2at} \tag{4.18}$$

$$\kappa_3(t) = np(1-p)(1-2p) + \lambda = \sigma^2(t) - 2X_0 e^{-2at}(1 - e^{-at}) \tag{4.19}$$

These analytical solutions are used subsequently to derive several properties of this model.

ii) Differential equations for any desired cumulants may be obtained by a series expansion of (4.16). Substituting definition (3.26) into (4.16) and equating corresponding coefficients of powers of θ, one finds

$$\dot{\kappa}_1(t) = I - a\kappa_1(t) \tag{4.20}$$

$$\dot{\kappa}_2(t) = I + a\kappa_1(t) - 2a\kappa_2(t) \tag{4.21}$$

$$\dot{\kappa}_3(t) = I - a\kappa_1(t) + 3a\kappa_2(t) - 3a\kappa_3(t) \tag{4.22}$$

The solutions to these equations, with $\kappa_1(0) = X_0$ and $\kappa_2(0) = \kappa_3(0) = 0$, are given in (4.17)–(4.19).

iii) The Kolmogorov equations in (4.6) may be solved numerically for the $p_i(t)$, as indicated in Section 4.3. The cumulant functions may then be obtained numerically using the definitions:

$$\begin{aligned}\kappa_1(t) &= \Sigma i p_i(t)\\ \kappa_2(t) &= \Sigma(i-\kappa_1(t))^2 p_i(t)\\ \kappa_3(t) &= \Sigma(i-\kappa_1(t))^3 p_i(t)\end{aligned} \tag{4.23}$$

The first approach is elegant, but is difficult to apply to other, more general population models. The second approach is very general and usually relatively easy to implement; it is therefore used extensively throughout this monograph. The third approach is useful, particularly for subsequent models, when the first two are not available. However this approach for the LID model with $a > 0$ has the drawback that the $p_i(t)$ numerical solutions are only approximate. Hence the $\kappa_i(t)$ from (4.23) would also be approximate, albeit very close approximations.

A number of variations are possible. One is the direct differentiation of the generating function $K(\theta, t)$ in (4.15) or of $P(s,t)$ in (4.10) using relationships such as (3.27) and (3.20) directly. This method is relatively easy to apply to the LID model, but again is not practical for more general models. Other small variations of these approaches, including the use of moment generating functions, may be helpful for specialized circumstances but will not be pursued either.

4.6 Some Properties of the Stochastic Solution

The cumulants for the LID model have several properties of interest. One is that "the solutions for the mean value for (linear) stochastic systems are the same as the solutions for the corresponding deterministic systems." [28, p. 236] As claimed, the mean value function in (4.17) is identical to the deterministic solution for $X(t)$ in (4.3).

A second property apparent from (4.17)–(4.19) is that, for the LID model,

$$\kappa_1(t) \geq \kappa_2(t) \geq \kappa_3(t) \text{ for } t \geq 0 \tag{4.24}$$

Indeed, equality can hold only for $t = 0$ and for the limiting ($t \to \infty$) Poisson distribution of X^*, in which case all cumulants are

$$\kappa_i^* = I/a, \quad i = 1, 2, \ldots \tag{4.25}$$

This property is often important in practice to identify the LID model. The coefficient of variation, i.e.

$$cv = (\kappa_2(t))^{1/2}/\kappa_1(t), \tag{4.26}$$

may be very small for large $\kappa_1(t)$, as noted in [28, p. 243].

4.7 Illustrations

4.7.1 Application to Bioaccumulation of Mercury in Fish

The concentration-time model (4.5), obtained from the deterministic model (4.3), was previously fitted to the mercury bioaccumulation data. One could also immediately derive this regression model for the expected concentration-time curve from the mean value function in (4.17) for the stochastic model. Hence there does not appear to be any *practical* difference between assuming a deterministic or a stochastic conceptualization of the model in the present single-compartment application. However it will be pointed out subsequently that the stochastic framework provides an assumed *residence time distribution* of mercury particles in the fish. Specifically, it will be shown that the particles have an exponential distribution with parameter a^{-1}, consequently the mean and variance for the time a random particle spends in a fish are a^{-1} and a^{-2}, respectively. Such characterization provides useful insight into mercury dynamics in fish, particularly when the single compartment model of a fish is compared to multiple compartment alternatives.

4.7.2 Application to AHB Population Dynamics

The LID model may be applied to describe AHB population dynamics during periods when swarming (births) do not occur. In this application, the number of colonies (or generic "particles") may be small, and therefore the stochastic model will be fully developed in order to illustrate its richness.

Consider the following assumed rates for the AHB model:

$$I = 1.4 \text{ colonies /time}$$
$$a = 0.08 \text{ (time}^{-1}\text{)}.$$

These assumed rates provide a rough linear model approximation to a more realistic, nonlinear AHB model introduced in Chapter 6. Assuming also $X(0) = 2$, the deterministic solution from (4.3) is

$$X(t) = 17.5 - 15.5e^{-0.08t}. \tag{4.27}$$

The equilibrium solution is clearly the constant $X^* = 17.5$

Consider now the stochastic solution. One immediate qualitative difference is that the equilibrium population size, X^*, is now by definition a random variable. Its equilibrium distribution is the Poisson given in (4.8), which is illustrated in Figure 4.2. It follows that the equilibrium mean, variance and skewness for these parameters values are

$$\mu = \sigma^2 = \kappa_3 = \lambda = 17.5, \tag{4.28}$$

which may also be obtained directly from (4.17)–(4.19).

Figure 4.2. Poisson equilibrium distribution for the LID model with immigration rate $I = 1.4$ and death rate $a = 0.08$, with a saddlepoint approximation.

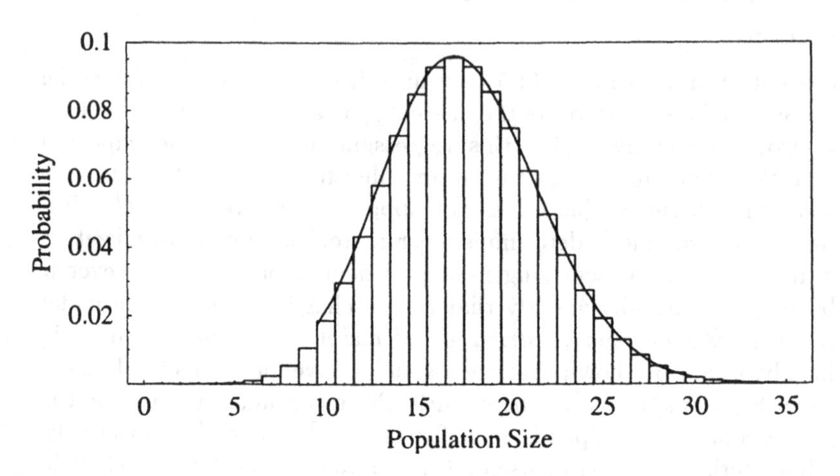

Figure 4.2 also illustrates the third-order saddlepoint approximation obtained by substituting the cumulants in (4.28) into (3.11). The approximation is admissible and very accurate in the range $X \geq 10$; for $X < 10$ higher order approximations would be required.

The transient probability functions for $X(t)$ may be obtained analytically from (4.11), but will not be pursued. Instead, we will obtain only the numerical solutions. Note that the coefficient matrix $\mathbf{R}$ in (4.6) for the Kolmogorov equations is

$$\mathbf{R} = \begin{bmatrix} -1.40 & 1.40 & 0 & 0 & \cdots \\ 0.08 & -1.48 & 1.40 & 0 & \cdots \\ 0 & 0.16 & -1.56 & 1.40 & \cdots \\ \vdots & & & & \end{bmatrix}.$$

Although $\mathbf{R}$ is infinite, accurate approximate solutions for the $p_i(t)$ probability functions were obtained by truncating the distribution at $U = 35$. The solution for $p_0(t)$, which of course could also be obtained directly from (4.14), is illustrated in Figure 4.3.

The cumulant functions could be obtained in the three ways mentioned in Section 4.5. Obviously, the direct solution from the analytical solutions in (4.17)–(4.19) is easily obtained. However, as an illustration, we note that they could also be obtained by solving the following set of linear differential equations from (4.20)–(4.22):

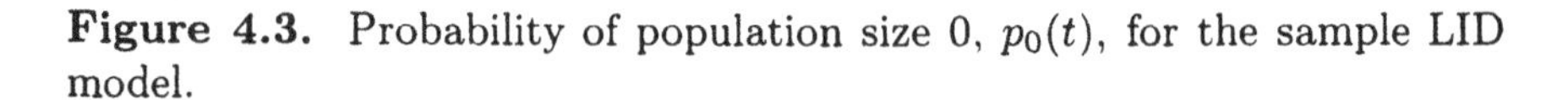

Figure 4.3. Probability of population size 0, $p_0(t)$, for the sample LID model.

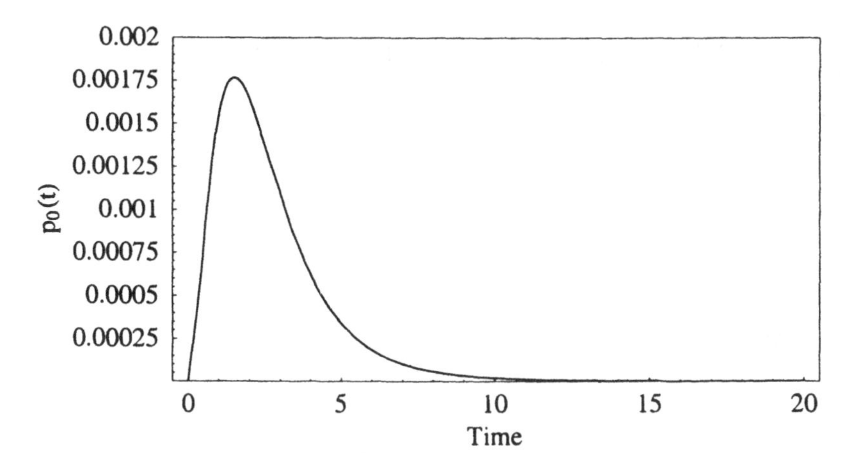

$$\begin{aligned}\dot{\kappa}_1\ (t) &= 1.4 - 0.08\kappa_1(t)\\ \dot{\kappa}_2\ (t) &= 1.4 + 0.08\kappa_1(t) - 0.16\kappa_2(t)\\ \dot{\kappa}_3\ (t) &= 1.4 - 0.08\kappa_1(t) + 0.24\kappa_2(t) - 0.24\kappa_3(t).\end{aligned}$$

The solutions are given in Figure 4.4. It is apparent that the equilibrium value is equal (i.e. 17.5) for all three cumulant functions. It is also clear from the graphs that property (4.24) holds.

Figure 4.4. First three cumulant functions, $\kappa_1(t)$, $\kappa_2(t)$, and $\kappa_3(t)$, for the sample LID model. In the graph, $\kappa_1(t) > \kappa_2(t) > \kappa_3(t)$ for any t.

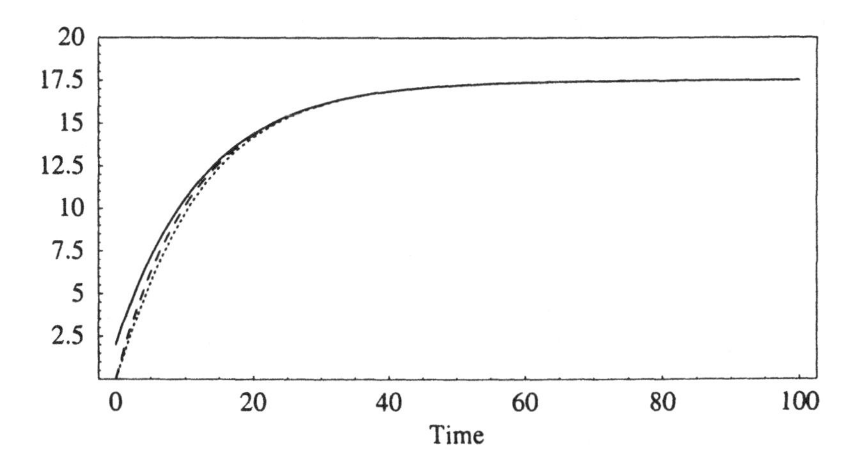

5

Linear Birth-Immigration-Death Models

5.1 Introduction

Consider now modeling the single population of size $X(t)$ with linear birth and death rates:

$$\lambda_X = a_1 X \quad \text{and} \quad \mu_X = a_2 X. \tag{5.1}$$

The immigration rate is again assumed to be I. This is called the LBID model. Clearly this model with linear (density-independent) growth rates could hold in the case of $a_1 > a_2$ only for periods of initial population growth. The model will be illustrated for such initial population growth for both the African bee and the muskrat examples in Chapter 2.

5.2 Deterministic Model

5.2.1 Solution to Deterministic Model

The deterministic model may be written, for subsequent convenience, as:

$$\dot{X}(t) = aX + I \tag{5.2}$$

where a is the net rate

$$a = a_1 - a_2. \tag{5.3}$$

The solution to (5.2) is

$$X(t) = X_0 e^{at} + (e^{at} - 1)I/a. \tag{5.4}$$

An inherent assumption in standard compartmental modeling is that $a < 0$, i.e. in the present context that the death rate exceeds the birth rate. In that case, the solution is equivalent to (4.3) and has equilibrium value $-I/a$. The more interesting case for the present population modeling, namely $a > 0$, is outside the scope of what is usually defined as compartmental modeling.

5.2.2 Application to AHB Population Dynamics

As an illustration, consider a LBID model with the following parameters for the deterministic model:

$$I = 1.4 \qquad a_1 = 0.08$$
$$X(0) = 2 \qquad a_2 = 0.16$$

With a negative net rate, $a = -0.08$, the deterministic solution from (5.4) for this LBID model is identical to that of the previous LID example given in (4.27). Recall that the asymptotic solution is $X^* = 17.5$.

5.3 Probability Distributions for the Stochastic Model

5.3.1 Kolmogorov Equations

Though the addition of the linear birth mechanism does not change the deterministic model, the same is certainly not true for the stochastic model. Consider first finding the transient probability functions. The Kolmogorov forward equations for the LBID model are

$$\dot{p}_x(t) = \begin{cases} -Ip_0 + a_2 p_1 & \text{for } x = 0 \\ [I + a_1(x-1)]p_{x-1} - [I + (a_1 + a_2)x]p_x & \\ \qquad + a_2(x+1)p_{x+1} & \text{for } x > 0 \end{cases} \tag{5.5}$$

The coefficient matrix $\mathbf{R}$ for the system of equations is again tridiagonal, with elements

$$r_{i,j} = \begin{cases} r_{i,i+1} = I + ia_1 & \\ r_{i,i-1} = ia_2 & \\ r_{i,i} = -I - i(a_1 + a_2) & \\ r_{i,j} = 0 & \text{otherwise} \end{cases} \tag{5.6}$$

In the case $a < 0$, one could again truncate the distribution of $X(t)$ at some large upper value, U, and solve the resulting finite system of equations given in (5.6) for numerical approximations of the $p_x(t)$ probability functions. For $a > 0$, such approximations could be obtained only for initial periods of elapsed time t.

5.3.2 Equilibrium Distribution

The equilibrium distribution for this model may be derived directly from the Kolmogorov equations [82, p.43], as previously illustrated for the LID model in Section 4.3. Setting $\dot{\mathbf{p}}(t) = \mathbf{0}$, with π_i again denoting $p_i(\infty)$, the equations in (5.5) give

$$\pi_1 = \pi_0(I/a_1)$$
$$\pi_2 = \pi_0 I(I + a_1)/2a_2^2$$

and in general, after some algebra,

$$\pi_i = \pi_0 (a_1/a_2)^i \binom{i - 1 + (I/a_1)}{i}. \tag{5.7}$$

Upon summing these up, one can solve for π_0, provided $a_1 < a_2$, i.e. $a < 0$ in (5.3), as

$$\pi_0 = (-a/a_2)^{I/a_1}$$

Substitution into (5.7) shows that X^* follows the negative binomial distribution, i.e.

$$\pi_i = \binom{k - 1 + i}{i} p^k (1 - p)^i$$

with $k = I/a_1$ and $p = -a/a_2$. Symbolically,

$$X^* \sim NB(k = I/a_1, p = -a/a_2). \tag{5.8}$$

This result will be proven using generating functions in the following section.

5.4 Generating Functions

The partial differential equation for the probability generating function of this LBID model is easily obtained using the operator equation in (3.32). The intensity functions are

$$f_1 = I + a_1 x \quad \text{and} \quad f_{-1} = a_2 x$$

which yield equation

$$\frac{\partial P}{\partial t} = [a_1 s(s - 1) + a_2(1 - s)] \frac{\partial P}{\partial s} + I(s - 1)P. \tag{5.9}$$

The analytical solution, given in a slightly different form in [4, p. 99], is:

$$P(s, t) = \frac{(a)^{I/a_1} \left\{a_2(e^{at} - 1) - (a_2 e^{at} - a_1)s\right\}^{X_0}}{\left\{(a_1 e^{at} - a_2) - a_1 s(e^{at} - 1)\right\}^{X_0 + I/a_1}} \tag{5.10}$$

where $a = a_1 - a_2$, as defined in (5.3).

Clearly, it would be very difficult to solve analytically for the $p_i(t)$ transient probability function for any $i > 0$ using successive differentiation in (3.20). However the probability function $p_0(t)$ may be obtained as $P(0, t)$ in (5.10), from whence

$$p_0(t) = a^{I/a_1} \left(a_2 e^{at} - a_2\right)^{X_0} \left(a_1 e^{at} - a_2\right)^{-X_0 - I/a_1}. \tag{5.11}$$

The equilibrium probability for the case $a < 0$ is

$$p_0(\infty) = (-a/a_2)^{I/a_1}. \tag{5.12}$$

Two special cases of the pgf are instructive. First, in the case of zero initial population size, i.e. $X(0) = 0$, (5.10) reduces to

$$P(s,t) = \left\{a_1 e^{at} - a_2 - a_1 s(e^{at} - 1)/a\right\}^{-I/a_1} . \tag{5.13}$$

The form of the pgf implies that $X(t)$ has a negative binomial distribution [30], with parameters

$$k = I/a_1 \quad \text{and} \quad p = a/(a_1 e^{at} - a_2). \tag{5.14}$$

Specifically, the probability density function is

$$\text{Prob}[X(t) = i] = \binom{k-1+i}{i} p^k (1-p)^i \tag{5.15}$$

with p and k as in (5.14).

The second special case assumes a negative net growth rate, i.e. $a < 0$. The limiting pgf for X^* for this case may be obtained from (5.10) as:

$$P(s,\infty) = [(a_1 s - a_2)/a]^{-I/a_1} . \tag{5.16}$$

The form of the pgf implies that the equilibrium distribution of $X(t)$ is a negative binomial, independent of X_0, with parameters k and $p = -a/a_2$ from (5.14), as previously indicated in Section 5.3.2. Substituting these equilibrium parameters into (5.15) yields $p_0(\infty)$ in (5.12).

The pde for the cumulant generating function, which may be obtained either by transforming (5.9) or by transforming the equation for the mgf from (3.33), is

$$\frac{\partial K}{\partial t} = \left\{a_1\left(e^{\theta} - 1\right) + a_2\left(e^{-\theta} - 1\right)\right\}\frac{\partial K}{\partial \theta} + I\left(e^{\theta} - 1\right). \tag{5.17}$$

This will be expanded subsequently to obtain equations for cumulants.

5.5 Cumulant Functions

Consider finding the cumulant functions for the LBID model using the three approaches described previously in Section 4.5 for the LID model. The first approach, which relies on the known cumulant structure for certain well-known distributions, has some limited application. Specifically, in the case $X(0) = 0$, $X(t)$ was found in (5.13) to have a negative binomial distribution. The first three cumulants for such distributions with parameters k and p are [30]:

$$\begin{aligned} \kappa_1 &= k(1-p)/p \\ \kappa_2 &= k(1-p)/p^2 \\ \kappa_3 &= k(1-p)(2-p)/p^3. \end{aligned} \tag{5.18}$$

Substituting the parameters from (5.14), it follows that the cumulant functions for this special case of $X(0) = 0$ are

$$\begin{aligned} \kappa_1(t) &= \left(e^{at} - 1\right) I/a \\ \kappa_2(t) &= \kappa_1(t) \left[\left(a_1 e^{at} - a_2\right)/a\right] \\ \kappa_3(t) &= \kappa_2(t) \left[\left(2a_1 e^{at} - a_1 - a_2\right)/a\right] \end{aligned} \tag{5.19}$$

The second approach consists of substituting the series expansion (3.26) into (5.17). We used symbolic math software [104] to obtain the equations, however they could readily be obtained directly without such software. The equations for the first three cumulant functions are:

$$\begin{aligned} \dot{\kappa}_1\ (t) &= I + a\kappa_1 \\ \dot{\kappa}_2\ (t) &= I + c\kappa_1 + 2a\kappa_2. \\ \dot{\kappa}_3\ (t) &= I + a\kappa_1 + 3c\kappa_2 + 3a\kappa_3 \end{aligned} \tag{5.20}$$

with, extending (5.3),

$$\begin{aligned} a &= a_1 - a_2 \\ c &= a_1 + a_2. \end{aligned} \tag{5.21}$$

The differential equations reduce to (4.17)–(4.19) for $a_1 = 0$.

Equations (5.19) could be solved recursively for the exact solution to these functions. The analytical solutions with $X(0) = X_0$ are:

$$\kappa_1(t) = X_0 e^{at} + (e^{at} - 1)I/a \tag{5.22}$$

$$\kappa_2(t) = X_0 c e^{at}(e^{at} - 1)/a + I(e^{at} - 1)(a_1 e^{at} - a_2)/a^2 \tag{5.23}$$

$$\begin{aligned} \kappa_3(t) = {} & X_0 e^{at} \left[3c^2(e^{at} - 1)^2 + a^2(e^{2at} - 1)\right] /2a^2 \\ & + I\left[-2ca_2 + (3c^2 - a^2)e^{at} - 6a_1 c e^{2at} + 4a_1^2 e^{3at}\right] /2a^3, \end{aligned} \tag{5.24}$$

which generalize (4.14)–(4.16).

The third approach is based on the numerical solutions of the $p_i(t)$ functions, which for the LBID model are only approximate. Hence the resulting numerical solutions for the cumulant functions would also be approximate. Therefore, even though the approximations are very accurate in practice, the third approach will not be pursued at present.

The LBID model illustrates a surprising phenomenon for this and for some subsequent models. The probability distribution for $X(t)$ is for all practical purposes analytically intractable for the general case of $X(0) \neq 0$, yet the low order cumulant functions have relatively simple analytical solutions in (5.22)–(5.24). Furthermore, only approximate numerical solutions are available for the $p_i(t)$ probability functions, whereas exact solutions may be obtained for the $\kappa_i(t)$ cumulant functions. These facts support the practical utility of using cumulant functions to study the stochastic solutions to the population models.

5.6 Some Properties of the Stochastic Solution

The mean, $\kappa_1(t)$, in (5.22) is again identical to the deterministic model, $X(t)$, in (5.4) by virtue of this being a linear kinetic system. One can show in (5.14) that $0 \leq p \leq 1$ for any a. Hence it follows from (5.18) in the two negative binomial cases, i.e. for $X(0) = 0$ and for the limiting distribution of X^*, that

$$\kappa_1(t) \leq \kappa_2(t) \leq \kappa_3(t).$$

This differs from a fundamental property in (4.24) for the previous LID model. Though the property does not hold for initial t in the case $X(0) \neq 0$, the fact that it holds for large values of t may be useful in model discrimination.

5.7 Illustrations

5.7.1 Application to AHB Population Dynamics

The AHB illustration in Section 5.2.2 demonstrated that the deterministic model depends only on the net birth (or death) rate, and therefore the deterministic solutions were identical for the LID and the LBID models with the same net rates. However, the stochastic solutions for these two models are very different, as previously indicated.

One simple, striking illustration of these differences is in their equilibrium distributions. The equilibrium distribution for the LBID example is a negative binomial with parameter values from (5.14) of $k = 17.5$ and $p = 0.5$, as is illustrated in Figure 5.1. The equilibrium mean obtained from (5.18) is $\kappa_1 = 17.5$, which is equal to the mean of the previous LID model. However the variance from (5.18) for this model is $\kappa_2 = 35$, which is twice the size of that for the corresponding LID model. This wider dispersion is apparent in the longer tail in Figure 5.1 as compared to that in Figure 4.2. The figure also illustrates the third-order saddlepoint approximation in the admissible range ($X^* \geq 14$).

The transient distribution functions may again be obtained from the Kolmogorov equations. The $\mathbf{R}$ matrix from (5.6) for these parameters has form:

$$\mathbf{R} = \begin{bmatrix} -1.40 & 1.40 & 0 & 0 & \dots \\ 0.08 & -1.64 & 1.56 & 0 & \dots \\ 0 & 0.16 & -1.88 & 1.72 & \dots \\ \vdots & & & & \end{bmatrix}.$$

Figure 5.1. Negative binomial equilibrium distribution for the LBID model with immigration rate $I = 1.4$, birth rate $a_1 = 0.08$, and death rate $a_2 = 0.16$, together with a saddlepoint approximation.

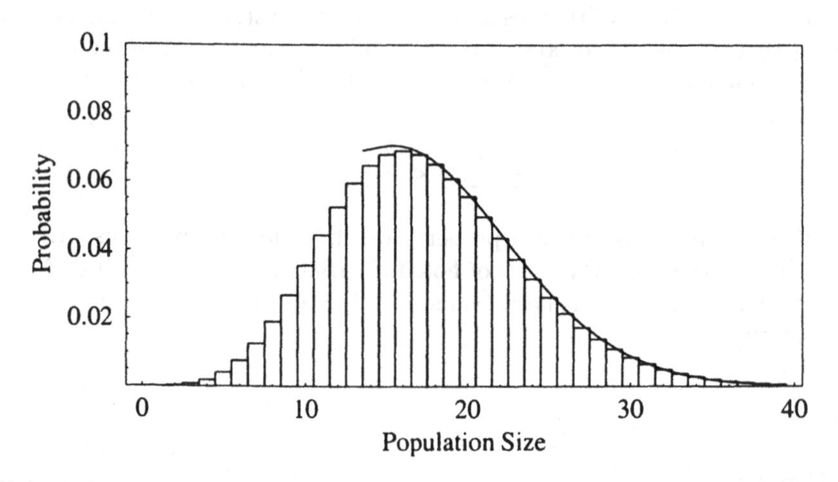

Accurate approximate solutions were obtained using $U = 50$ as a truncation value. The solution for $p_0(t)$ is given in Figure 5.2 as an illustration of these solutions. Clearly the exact solution for $p_0(t)$ could also be obtained directly from the analytical expression in (5.11).

The first three cumulant functions are available from (5.22)–(5.24) and are illustrated in Figure 5.3. Note that both the variance and skewness exceed the mean after a short initial transient period, as discussed in Section 5.6. The results from this section will be compared more extensively with those from the LID and other models in the Appendix of Chapter 7.

5.7.2 Application to Muskrat Spread

Consider now the data illustrated in Figure 2.4 for muskrat population size during their initial growth periods. One issue with these data is that they represent observed catch, say $Y(t)$, rather than total population size, $X(t)$. For the present model, this issue may be resolved by introducing a harvesting rate, say $h(t)$, which could also be regarded as the capture probability. Suppose that this rate is time invariant, say some constant h, which is a practical assumption for data analysis. The catch is then assumed to be

$$Y(t) = hX(t). \tag{5.25}$$

The regression model in (5.22) would thus require a change of scale, similar to the transformation to concentration data in Section 4.2.2. Letting $\kappa_1^{\dagger}(t)$

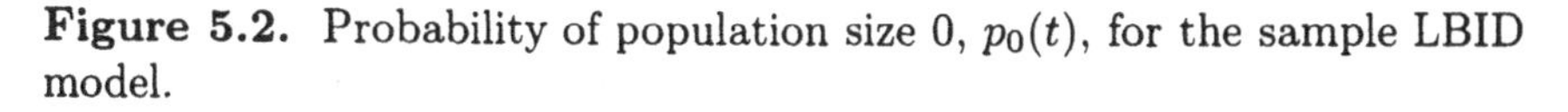

Figure 5.2. Probability of population size 0, $p_0(t)$, for the sample LBID model.

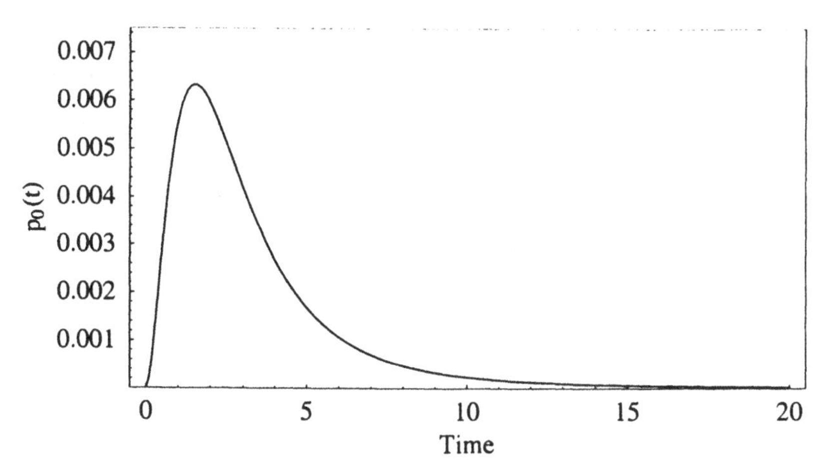

denote the mean of $Y(t)$, the model in (5.22) would be transformed to

$$\kappa_1^{\dagger}(t) = Y_0 e^{at} + (e^{at} - 1)I^{\dagger}/a, \tag{5.26}$$

where, as in (5.21), a denotes the net rate parameter which is unaffected by the transformation, Y_0 is the assumed catch at $t = 0$, and $I^{\dagger} = hI$. This model was fitted to the data for each of the 11 provinces in [53].

Detailed numerical results will be given for only four adjacent provinces, namely gelderl, overijl, dr, and gron, the four easternmost provinces in Figure 2.3. The gelderl data set started in 1968 and the others in 1970. Y_0 was set at 0 at each of these respective starting times, thus simplifying the model. The estimated net birth rates, $a = a_1 - a_2$, with standard errors in parentheses, are 0.365 (0.016), 0.296 (0.035), 0.421 (0.055) and 0.258 (0.023)/yr, respectively. The corresponding estimated immigration rates, $I^{\dagger}$, are 92 (16), 173 (49), 115 (46) and 342 (60) muskrats/yr. The fitted curves for these four provinces are illlustrated in Figure 5.4. These curves could be scaled up by the factor h^{-1} to estimate the deterministic growth curve in (5.4) or, of course, the equivalent mean function in (5.22) of population size $X(t)$ for the stochastic LBID model.

Estimation of the variance function in (5.23), however, requires estimates of both a_1 and a_2, which are not separately identifiable in the regression model of (5.4), (5.22) or (5.26). If data on population variances were available, say from sub-province geographical areas, the variance function in (5.23) could be fitted simultaneously with the mean in (5.22) to estimate a_1 and a_2 separately. Such data are not available in this example.

However, one extension of the problem is to incorporate exogenous information. The life expectancy of a muskrat is estimated to be about four years, hence we assume that $a_2 = 0.25$/yr and, using (5.3), find

$a_1 = (a + 0.25)/\text{yr}$. From these estimates, it is possible to solve for the variance and skewness functions in (5.23) and (5.24) for this assumed LBID model.

Figure 5.3. First three cumulant functions, $\kappa_1(t)$, $\kappa_2(t)$, and $\kappa_3(t)$, for sample the LBID model with $X(0) = 2$.

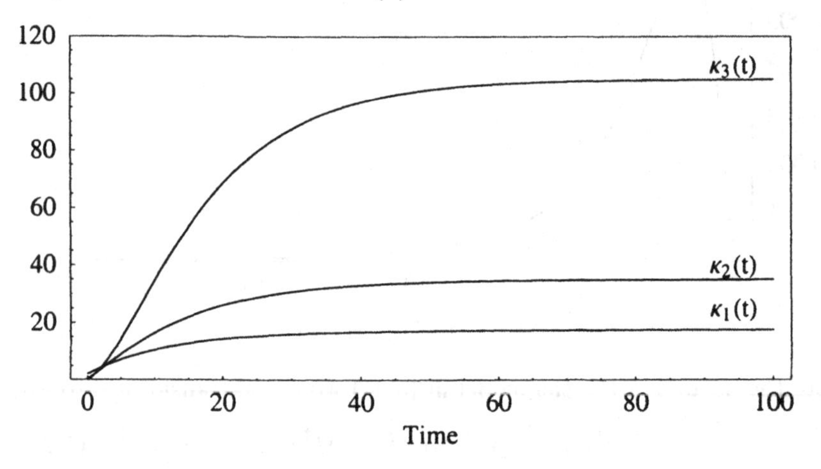

Figure 5.4. Observed muskrat catch data and fitted curves from LBID model for 4 provinces in the Netherlands.

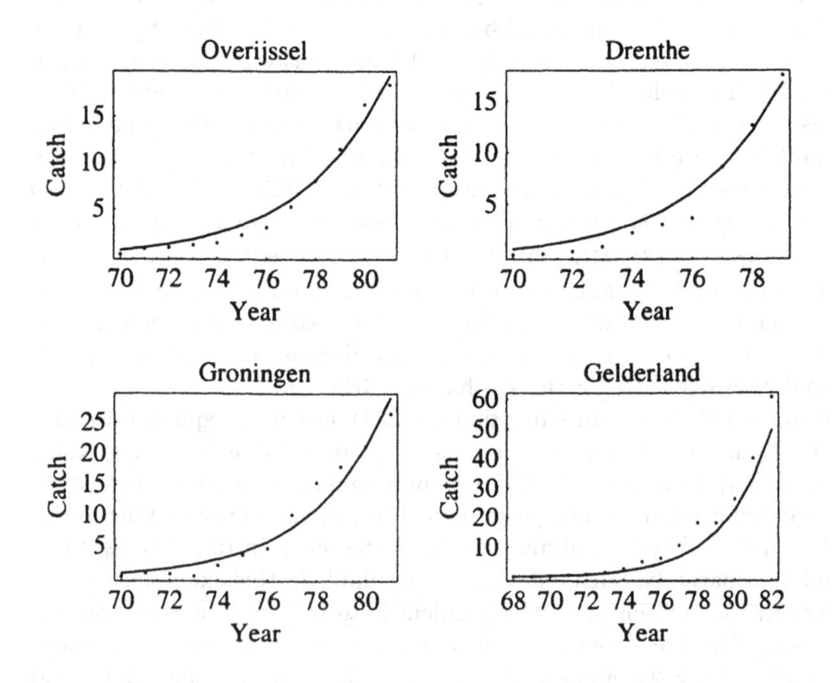

6

Nonlinear Birth-Death Models

6.1 Introduction

Consider now modeling the single population of size $X(t)$ with population rates

$$\lambda_X = \begin{cases} a_1 X - b_1 X^{s+1} & \text{for } X < (a_1/b_1)^{1/s} \\ 0 & \text{otherwise} \end{cases} \tag{6.1}$$

$$\mu_X = a_2 X + b_2 X^{s+1} \tag{6.2}$$

for integer $s \geq 1$. These are called nonlinear rates in ecological population modeling because the per capita rates, i.e. λ_X/X, are obviously functions of X, as discussed in Section 3.7. The a_i are called the intrinsic rates, and are interpreted as the per-capita birth and death rate coefficients for small initial population sizes, before population density pressures are of any practical consequence. The b_i are the crowding coefficients which add density-dependency to the model. Their inclusion may in principle yield long-term, equilibrium solutions for the model.

For simplicity, we assume in this chapter that there is no immigration, i.e. $I = 0$, and call this the NBD model. The model with immigration is considered in Chapter 7. The special case of $s = 1$, i.e. with quadratic birth and death rate functions, gives the classic Verhulst-Pearl model. It is widely studied and leads to the (ordinary) logistic model [80, 82]. The case of $s > 1$ is often called the power-law logistic model [6]. For simplicity, the same power s is used for both the birth and death rate functions in the power-law logistic model in (6.1) and (6.2). The subsequent theory, however, would hold for any general polynomial functions in (6.1) and (6.2).

This chapter follows the approach of the previous two chapters, in first deriving the deterministic model and then developing the stochastic model as outlined in Chapter 3. There are a number of specialized procedures for the NBD model, however, which may be helpful in simplifying computations and in contributing added insight into the dynamics of the population.

These specialized procedures will be presented in the appendix to this chapter.

6.2 Deterministic Model

6.2.1 Solution for Deterministic Model

The deterministic model with (6.1) and (6.2) may be written as

$$\dot{X}(t) = aX - bX^{s+1}. \tag{6.3}$$

where, extending again (5.3),

$$a = a_1 - a_2, \text{ and} \tag{6.4}$$
$$b = b_1 + b_2.$$

The model has solution [6, p. 108]:

$$X(t) = \frac{K}{[1 + m\exp(-ast)]^{1/s}} \tag{6.5}$$

with

$$K = (a/b)^{1/s} \text{ and } m = (K/X_0)^s - 1. \tag{6.6}$$

That K in (6.6) is the equilibrium population size is easy to show by setting $\dot{X}(t) = 0$ in (6.3) and solving for the equilibrium value of X. K is called the "carrying capacity" in ecological modeling.

The point of inflection for the model occurs at time $t_i = [\log_e(m/s)]/as$ with population size

$$X(t_i) = K/(1+s)^{1/s}. \tag{6.7}$$

Note that in the special case of the ordinary logistic growth curve with $s = 1$, the point of inflection occurs at population size $K/2$. This is an important property which may be useful in model discrimination. The ordinary logistic growth curve is widely used, however there is evidence in some applications, and in particular for both the AHB and the muskrat examples, that maximal growth occurs at population sizes substantially larger than $K/2$. For such applications, the models with $s > 1$ are obviously of great interest.

6.2.2 Application of Ordinary Logistic Model to AHB Dynamics

For present simplicity, we consider our standard single population AHB model, developed in [44]. This is an ordinary logistic ($s = 1$) model with parameter values:

$$a_1 = 0.30 \qquad a_2 = 0.02 \tag{6.8}$$
$$b_1 = 0.015 \qquad b_2 = 0.001.$$

To compare the model with previous linear illustrations, we assume $X(0) = 2$. The deterministic solution from (6.5) is

$$X(t) = 17.5[1 + 7.75\exp(-0.28t)]^{-1}.$$

This solution is plotted in Figure 6.1. The shape, though clearly different from the simple linear kinetic examples in Section 4.7.2 and 5.2.2, is similar to those from the linear examples, and will be compared in some detail subsequently in Appendix 7.8.2 of Chapter 7.

Figure 6.1. Deterministic solution, $X(t)$, and mean value function, $\kappa_1(t)$, for the stochastic solution to the NBD (logistic) model with initial size $X(0) = 2$ and with standard AHB parameters $a_1 = 0.30$, $a_2 = 0.02$, $b_1 = 0.015$ and $b_2 = 0.001$.

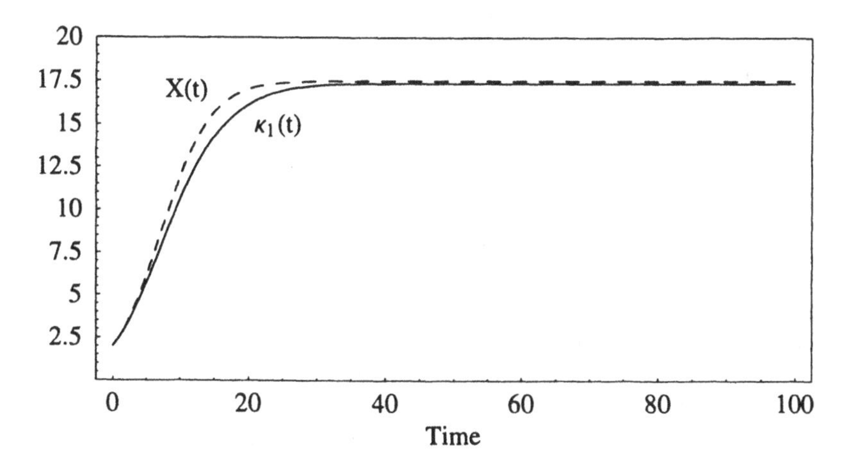

6.3 Probability Distributions for the Stochastic Model

6.3.1 Kolmogorov Equations

Let

$$u = (a_1/b_1)^{1/s} \tag{6.9}$$

in (6.1). It is clear that the birthrate is 0 for $X > u$, and hence that the population size is bounded above by u. For simplicity, we assume that u in an integer; the more general case where u is not restricted to be integer-valued is considered in [44].

The Kolmogorov forward equations for the NBD model are:

$$\begin{aligned}
\dot{p}_0\ (t) &= \mu_1 p_1(t) \\
\dot{p}_1\ (t) &= -(\lambda_1 + \mu_1)p_1(t) + \mu_2 p_2(t) \\
\dot{p}_x\ (t) &= \lambda_{x-1}p_{x-1}(t) - (\mu_x + \lambda_x)p_x(t) + \mu_{x+1}p_{x+1}(t), \\
&\qquad \text{for } x = 2, \ldots, u-1, \\
\dot{p}_u\ (t) &= \lambda_{u-1}p_{u-1}(t) - \mu_u p_u(t).
\end{aligned} \tag{6.10}$$

The corresponding tridiagonal coefficient matrix **R** has elements

$$r_{i,j} = \begin{cases} r_{oj} = 0 & \text{for } 0 \leq j \leq u \\ r_{i,i+1} = \lambda_i & \text{for } 0 < i < u \\ r_{i+1,i} = u_i & \text{for } 0 \leq i < u \\ r_{i,i} = -(\lambda_i + u_i) & \text{for } 0 < i < u \end{cases} \tag{6.11}$$

Because the system of equations is finite, one may obtain exact numerical solutions for these $p_x(t)$ probability functions. We used general numerical techniques software, however new computational methods are also available [77]. The fact that exact solutions are available may account in large measure for the relatively widespread use of this model, as compared to the subsequent nonlinear model with immigration where only approximate solutions exist.

6.3.2 Quasi-Equilibrium Population Size Distribution

In the absence of immigration, it is clear from (6.1) that population size 0 is an absorbing state and that hence $p_0(t)$ is the "extinction" probability by time t. Moreover, because the population size is bounded above by u in (6.9), it follows that ultimate extinction is *certain*, i.e. $p_0(\infty) = 1$. The exact solution for $p_0(t)$ may be obtained numerically from (6.10), as previously noted.

Bartlett *et al.* [7] observe that, although extinction is certain, such extinction may not occur within realizable time intervals. A population is defined to be *ecologically stable* if it persists for a long period, otherwise it is ecologically unstable [82]. Let E_x denote the mean time to extinction. The computation of E_x is discussed in Appendix 6.8. The *stability index* is defined as

$$\xi_x = \log E_x. \tag{6.12}$$

The concept of a quasi-equilibrium distribution, which by definition is conditioned on $X(t) > 0$, has been developed for populations which are ecologically stable with large stability indices. The computation of such quasi-equilibrium distributions is based on the concept that a population in equilibrium would not "drift" [7]. Accordingly, the probability of the joint event of the population being of size x *and* having a loss (or "death") of a single unit equals the probability of the population being of size $x-1$ *and* having an increase (or "birth") of a single unit. In symbols, the probabilities would satisfy the relationship

$$\mu_x p_x(t) = \lambda_{x-1} p_{x-1}(t) \tag{6.13}$$

for $1 < x \leq u$.

The quasi-equilibrium distribution, with π_x denoting $p_x(\infty)$, may be calculated numerically from the recursion relationship above. The mechanics of solving for this distribution numerically are not difficult, and the computations are illustrated for the ordinary logistic model with $s = 1$ in many leading ecological modeling texts,(see e.g. [80, p. 32] and [82, p. 65]). This quasi-equilibrium distribution will be illustrated later in Chapter 6. Also, an alternative approach for obtaining this distribution is given in Appendix 6.8.2.

6.4 Generating Functions

The intensity functions for the assumed model are

$$\begin{aligned} f_1 &= a_1 x - b_1 x^{s+1} \\ f_{-1} &= a_2 x + b_2 x^{s+1} \end{aligned}$$

for $x \leq u$, which may be substituted into the operator equations to yield partial differential equations for generating functions. For example, the equation for the pgf from (3.32) for the case $s = 1$ is:

$$\frac{\partial P}{\partial t} = (s-1)(a_1 s - a_2)\frac{\partial P}{\partial s} + s(s-1)(b_1 s + b_2)\frac{\partial^2 P}{\partial s^2}. \tag{6.14}$$

The equation is analytically intractable.

The equation for the mgf from (3.33) for general s has the relatively simple form:

$$\begin{aligned} \frac{\partial M}{\partial t} &= \left[\left(e^{\theta}-1\right)a_1 + \left(e^{-\theta}-1\right)a_2\right]\frac{\partial M}{\partial \theta} \\ &\quad + \left[\left(e^{\theta}-1\right)(-b_1) + \left(e^{-\theta}-1\right)b_2\right]\frac{\partial^{s+1} M}{\partial \theta^{s+1}}. \end{aligned} \tag{6.15}$$

The corresponding equation for the cgf is found by substituting $K = \log M$ into (6.15). For $s = 1$, this transformation yields

$$\begin{aligned} \frac{\partial K}{\partial t} &= \left[(e^{\theta}-1)a_1 + (e^{-\theta}-1)a_2\right]\frac{\partial K}{\partial \theta} \\ &\quad + \left[(e^{\theta}-1)(-b_1) + (e^{-\theta}-1)b_2\right]\left[\frac{\partial^2 K}{\partial \theta^2} + \left(\frac{\partial K}{\partial \theta}\right)^2\right] \end{aligned} \tag{6.16}$$

As s increases the second term on the right becomes increasingly complicated but may be found using computer software.

Differential equations for the cumulant functions may again be obtained by series expansions of equations like (6.16), using (3.26), as illustrated in [56]. Some equations for the mean value, $\kappa_1(t)$, are:

$$\text{for } s = 1; \quad \dot{\kappa}_1(t) = (a - b\kappa_1)\kappa_1 - b\kappa_2 \tag{6.17}$$

$$\text{for } s = 2; \quad \dot{\kappa}_1(t) = \left(a - b\kappa_1^2\right)\kappa_1 - b\left(\kappa_3 + 3\kappa_1\kappa_2\right)$$

$$\text{for } s = 3; \quad \dot{\kappa}_1(t) = \left(a - b\kappa_1^3\right)\kappa_1 - b\left(\kappa_4 + 4\kappa_3\kappa_1 + 3\kappa_2^2 + 6\kappa_2\kappa_1^2\right)$$

Equations for the variance function, $\kappa_2(t)$, for some small s are:

$$\begin{aligned}
\text{for } s = 1; \quad \dot{\kappa}_2(t) &= (c - d\kappa_1)\kappa_1 + (2a - d - 4b\kappa_1)\kappa_2 - 2b\kappa_3 \\
\text{for } s = 2; \quad \dot{\kappa}_2(t) &= \left(c - d\kappa_1^2\right)\kappa_1 + \left(2a - 3d\kappa_1 - 6b\kappa_1^2 - 6b\kappa_2\right)\kappa_2 \\
&\qquad - (d + 6b\kappa_1)\kappa_3 - 2b\kappa_4, \qquad\qquad (6.18) \\
\text{for } s = 3; \quad \dot{\kappa}_2(t) &= \left(c - d\kappa_1^3\right)\kappa_1 + (2a + 6d\kappa_1^2 + 8b\kappa_1^3 + 3d\kappa_2 \\
&\qquad + 24b\kappa_1\kappa_2)\kappa_2 - 4\left(d\kappa_1 + 3b\kappa_1^2 + 5b\kappa_2\right)\kappa_3 \\
&\qquad - (d + 8b\kappa_1)\kappa_4 - 2b\kappa_5,
\end{aligned}$$

where, extending (6.4),

$$a = a_1 - a_2, \quad b = b_1 + b_2, \quad c = a_1 + a_2, \quad \text{and } d = b_1 - b_2. \tag{6.19}$$

In principle, one could derive directly such equations for any cumulant of any s-power model, but the algebra becomes tedious. Instead, expressions for higher order cumulants and for larger s may be obtained from (6.15) using symbolic computer software packages such as *Mathematica* [104].

For subsequent comparative purposes, the equation for the skewness function with $s = 1$ is:

$$\begin{aligned}
\dot{\kappa}_3(t) &= (a - b\kappa_1)\kappa_1 + (3c - b - 6d\kappa_1 - 6b\kappa_2)\kappa_2 \\
&\quad + (3a - 3d - 6b\kappa_1)\kappa_3 - 3b\kappa_4.
\end{aligned}$$

6.5 Cumulant Functions

Equations (6.17) and (6.18) suggest, and it is easily proven, that the differential equation for the jth cumulant function for an s-degree power law logistic model involves terms up to the $(j + s)^{th}$ cumulant. Obviously this fact rules out exact solutions, such as those previously found for the linear kinetic models, for the present equations. A standard approach to this problem has been to assume that the population size variable follows a Normal distribution, as in [103]. Consequently, incorporating properties of the Normal distribution [29], all cumulants of order 3 or higher are "neglected", i.e. set to 0.

We propose in [56], instead, finding approximating cumulant functions for this nonlinear kinetic model by using a "cumulant truncation" procedure. In this approach, one approximates the cumulant functions of any specific order, say i, of an s-power law model by solving a system of up to the first $(i+s)$ cumulant functions with all higher order cumulants set to 0. This would involve using cumulant functions beyond the second order in all except the simplest possible case. The accuracy of these approximate cumulant functions depends clearly on the specific parameter values, however our investigations concerning their accuracy have been very encouraging, as discussed in the illustrations to follow.

6.6 Some Properties of the Stochastic Solution

Two observations seem noteworthy. First, Nisbet and Gurney [73, Section 6.2] consider a comparable birth-death model and develop cumulant equations for the conditional probability distribution function (CPDF) where extinction is precluded (i.e. with $X(t) \geq 1$). The cumulant equations in (6.17) and (6.18) describe the (unconditional) distribution of $X(t)$ where extinction is not precluded, and hence would obviously be different.

Second, as Jacquez [28, p. 271] points out, the "means (of the stochastic system) no longer follow the same time course as the solution of the deterministic system." For the present model with $s = 1$, it is clear comparing (6.3) to (6.17) that $X(t)$ will exceed the corresponding $\kappa_1(t)$ for $t > 0$, because the differential equation for the latter contains a second term which is negative [44]. This property is of considerable interest, inasmuch as many modeling practitioners feel that the solutions for the deterministic model and for the mean of the stochastic model are indistinguishable in practice, if not identical. Instead, the difference between the two is an increasing function of the population variance, as illustrated subsequently.

6.7 Illustrations

6.7.1 Application of Ordinary Logistic Model to AHB Dynamics

Consider first the analogous stochastic model to the deterministic logistic model in Section 6.2.2. The coefficient matrix **R** from (6.11) for the stochastic model with parameters in (6.8) is

$$\mathbf{R} = \begin{bmatrix} 0 & 0 & 0 & 0 & \dots & 0 & 0 \\ .021 & -.306 & .285 & 0 & \dots & 0 & 0 \\ 0 & .044 & -.584 & .540 & \dots & 0 & 0 \\ \vdots & \vdots & \vdots & \vdots & \dots & \vdots & \vdots \\ 0 & 0 & 0 & 0 & \dots & -1.026 & .285 \\ 0 & 0 & 0 & 0 & \dots & .800 & -.800 \end{bmatrix}.$$

Note from (6.9) that $u = 20$.

The solution for the $p_0(t)$ function is illustrated in Figure 6.2. In the absence of immigration, $p_0(t)$ is an increasing function, thus differing qualitatively from the corresponding solutions in (4.14) and (5.11) for the previous (linear) models with immigration.

Figure 6.2. Probability of population size 0, $p_0(t)$, for the NBD model with $X(0) = 2$ and standard AHB parameters.

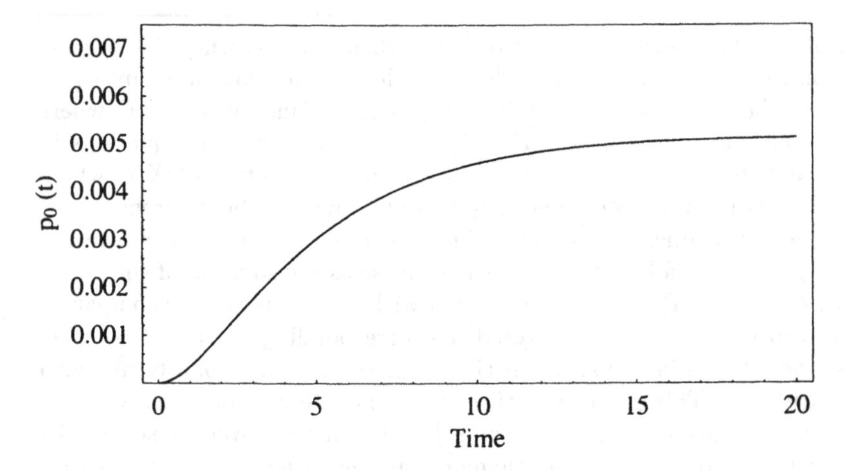

Though $p_0(t)$ is an increasing function, it is clear from Figure 6.2 that this extinction probability is very small over the present range of interest, i.e. $t < 100$. In fact, the mean time to extinction, E_1, for the given parameter values of this model is shown subsequently to be 6.1267×10^{12}. Therefore the model is ecologically stable and the assumption of a quasi-equilibrium distribution is reasonable. The numerical solution for the distribution derived from (6.13) is given in Figure 6.3.

The solutions for the cumulant functions obtained as described in Section 6.5 are illustrated in Figure 6.4. Figure 6.1 compares the deterministic solution, $X(t)$, to the corresponding mean function, $\kappa_1(t)$. The two curves differ substantially for some moderate t, though the equilibrium values are close with $X^* = 17.5$ and $\kappa_1(\infty) = 17.3564$.

Figure 6.3. Quasi-equilibrium population size distribution for the NBD model with standard AHB parameters.

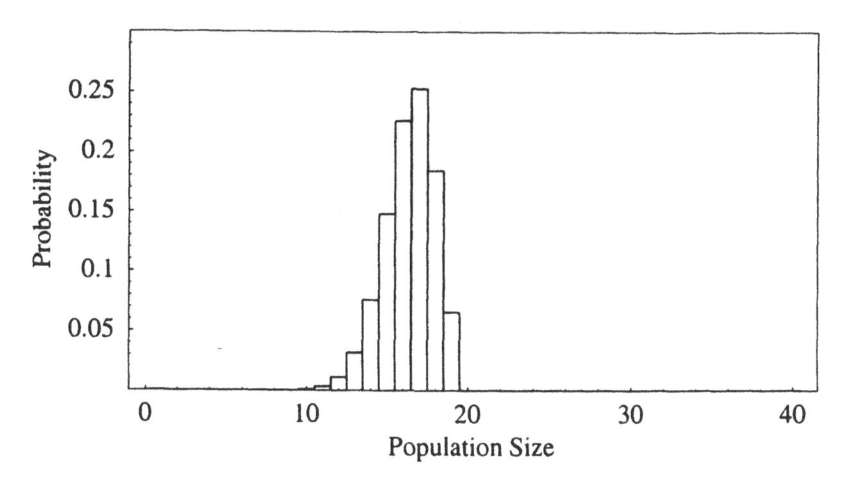

Figure 6.4. First three cumulant functions, $\kappa_1(t)$, $\kappa_2(t)$, and $\kappa_3(t)$, for the NBD model with $X(0) = 2$ and standard AHB parameters.

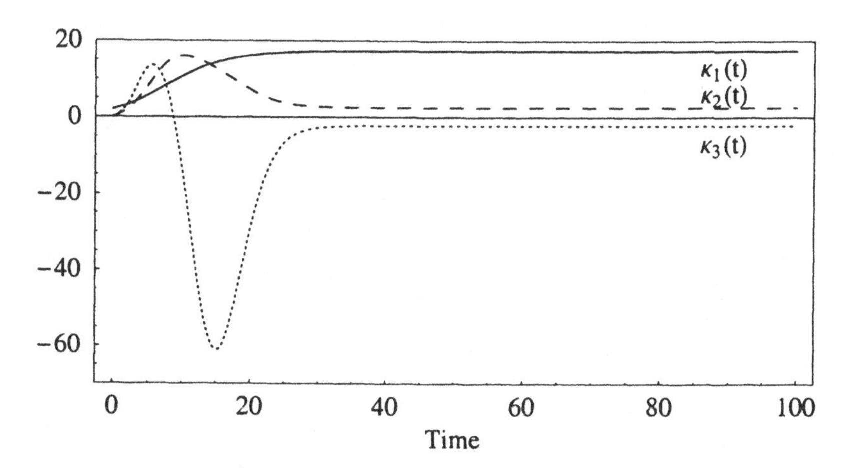

6.7.2 *Application of Power-Law Logistic Models to AHB Dynamics*

As noted in Chapter 2, the AHB has been labelled an "r-strategist" in its colonizing behavior because it is known to reproduce rapidly until very close to its carrying capacity. This suggests that the appropriate NBD growth model would have $s > 1$, because the size of such population at its point of inflection, i.e. at its time of maximal growth, would exceed the value of $K/2$ given in (6.7) for the deterministic model with $s = 1$.

A family of models with various small integer $s \geq 1$ in the rate functions (6.1) and (6.2) is investigated in [56] for use in describing AHB population growth. Our approach for obtaining families of models was first to fix the four parameters which seem most interpretable and estimable in practice, namely the intrinsic rates, a_1 and a_2, the carrying capacity, K, and the upper limit, u. The corresponding b_1 and b_2 parameter values for various s values could then be obtained from (6.6) and (6.9).

Consider now the specific family of models using the fixed parameters of the previous standard AHB model with $s = 1$ in Sections 6.2.2 and 6.7.1, i.e.

$$a_1 = 0.30 \qquad a_2 = 0.02$$
$$K = 17.5 \qquad u = 20.$$

The solutions for the b_1 and b_2 parameters for other models in this family with integer s are given in Table 6.1, and the corresponding birth and death rate functions are plotted in Figure 6.5.

Figure 6.6 illustrates the deterministic solutions for these four models with initial value $X(0) = 5$. The points of inflection from (6.7) occur at increasingly larger population sizes as s increases, as listed in Table 6.1. Such rapid and prolonged growth patterns are consistent with the known population dynamics of the AHB.

Figure 6.5. Assumed birth and death rate functions for family of postulated AHB population growth models, $s = 1$ to 4, with common parameters $a_1 = 0.30$, $a_2 = 0.02$, $K = 17.5$ and $u = 20$.

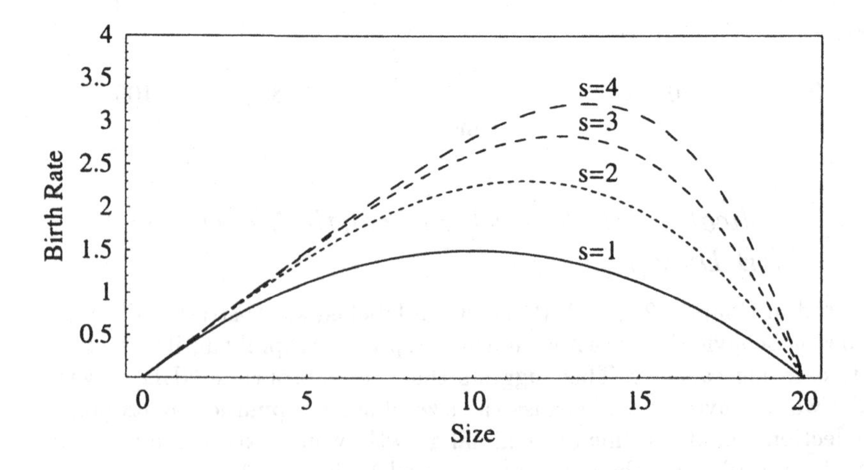

Figure 6.5. (Continued)

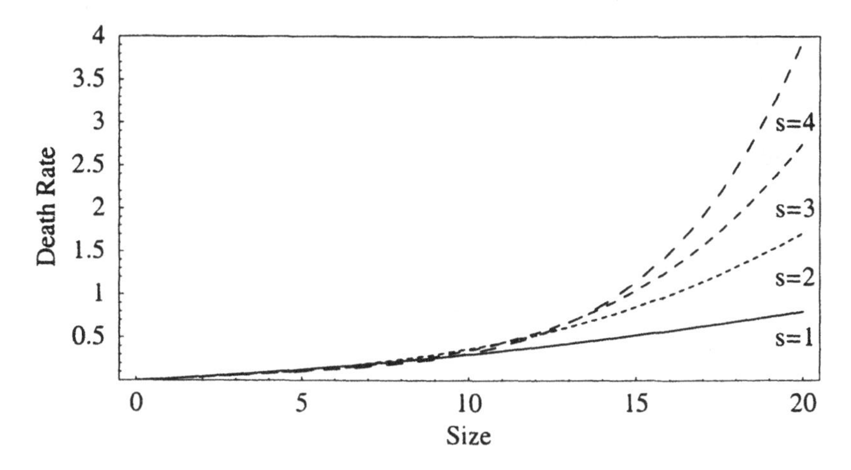

Figure 6.6. Solutions for population size function, $X(t)$, for the family of deterministic models, $s = 1$ to 4, for AHB model with initial size $X(0) = 5$.

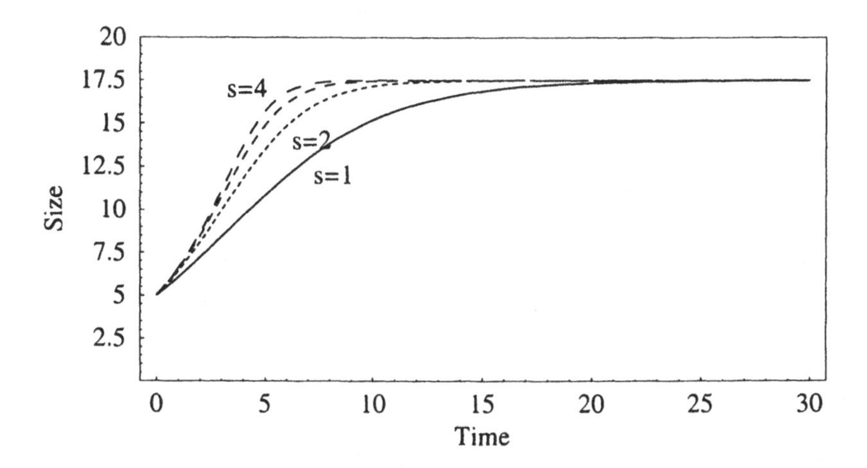

The complete solutions for the transient probability distributions of the stochastic models were obtained for each s by solving for the $p_x(t)$ functions in (6.10) with the appropriate $\mathbf{R}$ coefficient matrices and with initial size $X(0) = 5$. Figure 6.7 illustrates these exact solutions for the case $s = 1$; the corresponding solutions for $s = 2$ are given in [56]. The first three cumulant functions $\kappa_1(t)$, $\kappa_2(t)$, and $\kappa_3(t)$, were then calculated for each model from definitions (3.8) and (3.10).

Table 6.1. Numerical comparisons for a family of power law logistic models for $s = 1$ to 4. The common parameters are $a_1 = 0.30$, $a_2 = 0.02$, $K = 17.5$, and $u = 20$.

Power s	1	2	3	4
Coefficients:				
b_1	1.5E-2	7.50E-4	3.75E-5	1.875E-6
b_2	0.1E-2	1.643E-4	1.4745E-5	1.1104E-6
Inflection Points:				
det $N(t)$	8.75	10.10	11.02	11.70
stoch $\kappa_1(t)$	8.07	9.01	9.56	10.05
Maximum % Error				
$\widehat{\kappa}_1(t)$	0.007	0.015	0.013	0.018
$\widehat{\kappa}_2(t)$	0.750	1.810	1.410	1.250
$\widehat{\kappa}_3(t)$ extremes	2.100	5.600	1.900	2.200
Approx. Cumulants of Size Distribution at $t = 50$:				
$\widehat{\kappa}_1(50)$	17.3561	17.2997	17.2469	17.2028
$\widehat{\kappa}_2(50)$	2.4866	2.3804	2.2644	2.1480
$\widehat{\kappa}_3(50)$	-2.1099	-1.8387	-1.5171	-1.2447
Cumulants of Quasi-Equilibrium Size Distribution:				
$\kappa_1(\infty)$	17.3564	17.2997	17.2469	17.2028
$\kappa_2(\infty)$	2.4917	2.3804	2.2644	2.1480
$\kappa_3(\infty)$	-2.2101	-1.8388	-1.5169	-1.2448

These exact cumulant functions were compared for each model to the approximations, $\widehat{\kappa}_1(t)$, $\widehat{\kappa}_2(t)$, and $\widehat{\kappa}_3(t)$, obtained as outlined in Section 6.5 by truncating high order cumulants. These approximate solutions are illustrated in Figure 6.8 for the four models. The maximum percent error, $100|\widehat{\kappa}_i(t) - \kappa_i(t)|/\kappa_i(t)$, over the range $0 < t < 50$ was calculated for each of the four models and three cumulants, and these maximum error rates are listed in Table 6.1.

Outstanding among the findings is that these approximate cumulant functions are very accurate. The error in $\widehat{\kappa}_1(t)$ was less than 0.02% for each model over the range of t, whereas the error was less than 2% for $\widehat{\kappa}_2(t)$. The percent error over the full range of t is not meaningful for the skewness, $\widehat{\kappa}_3(t)$, because it crosses 0. However the errors at its maximum and minimum extremes were all less than 6%. In our investigations with other parameter values, the error rates for these cumulant approximations are typically even smaller.

Figure 6.7. Graphs of the probability functions $p_x(t)$; $x = 4, \ldots, 20$; for AHB population growth model with $s = 1$, initial size $X(0) = 5$ and standard rate parameters.

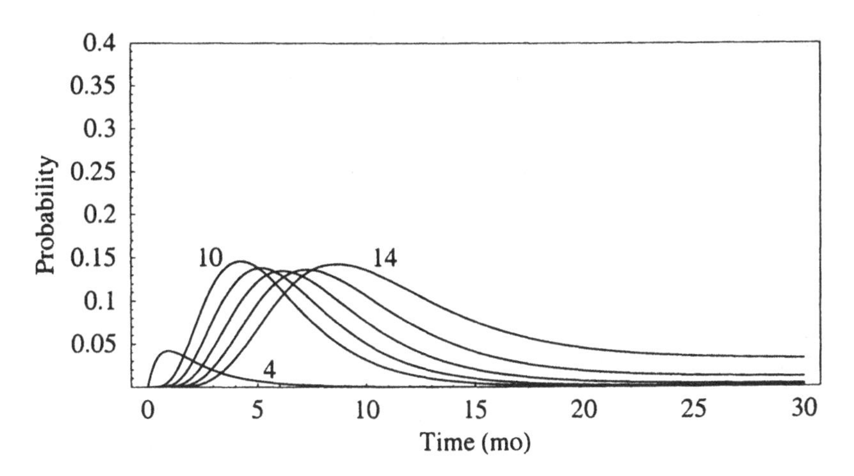

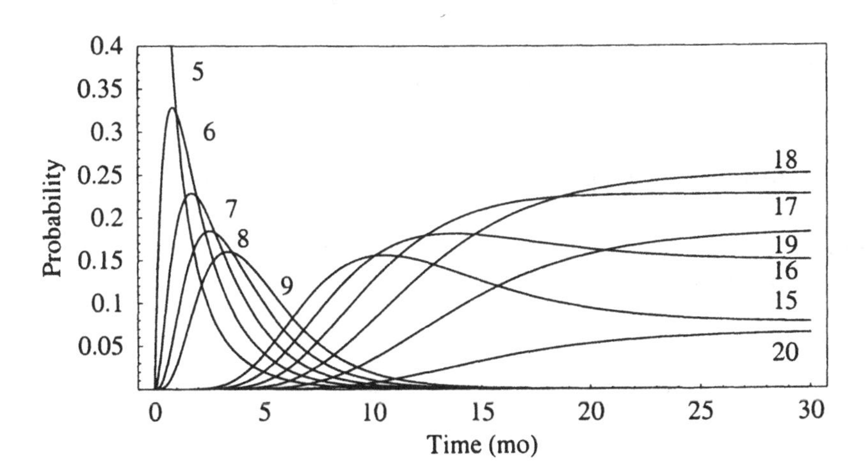

The approximate cumulant values for the population size distributions at $t = 50$ are listed in Table 6.1. It is apparent in Figure 6.8 that the cumulant functions stabilized long before $t = 50$. These values may be compared to the exact cumulants of the assumed quasi-equilibrium distributions, such as that illustrated in Figure 6.3 for the case $s = 1$. Table 6.1 lists these quasi-equilibrium cumulants for each model. The values for the means agree with six place accuracy, except for the $s = 1$ model with five place accuracy. The corresponding variances all agree with five place accuracy, except for the case with $s = 1$ where the relative error is less than 0.5%. The corresponding skewnesses agree with four place accuracy, again except for the $s = 1$ case where the relative error is less than 5%. These findings support the assertion that these AHB population models are ecologically stable with realistic quasi-equilibrium distributions.

As discussed in Section 6.6, the deterministic solutions for $X(t)$ in Figure 6.6 differ from the solutions for $\kappa_1(t)$ in Figure 6.8. One immediate difference apparent in Table 6.1 is that the deterministic carrying capacity, $K = 17.5$, exceeds the equilibrium mean value for each of the models. Another is that the points of inflection for $X(t)$, given by (6.7), are larger than those of $\kappa_1(t)$ for comparable s, as also apparent in Table 6.1. A related difference is that the approach to equilibrium is more gradual for $\kappa_1(t)$ than for $X(t)$. Of course, in a more global way, the fundamental difference between the deterministic and stochastic models for this AHB example is the richness of the population size distributions of the stochastic model, as illustrated in Figures 6.3 and 6.7, as compared to a deterministic size function, without uncertainty, as illustrated in Figure 6.1.

6.7.3 *Application to Muskrat Population Dynamics*

The muskrat populations in four of the eleven provinces, namely limb, gelderl, overijl and dr, appear to have leveled off in Figure 2.4 and Table 2.2 to some apparent quasi-equilibrium sizes prior to 1991, the last year of observation in the data set. For subsequent comparative purposes, we consider data from each of the latter three adjacent provinces, which were

Figure 6.8. Approximate mean, variance and skewnesss functions for the family of postulated AHB growth models.

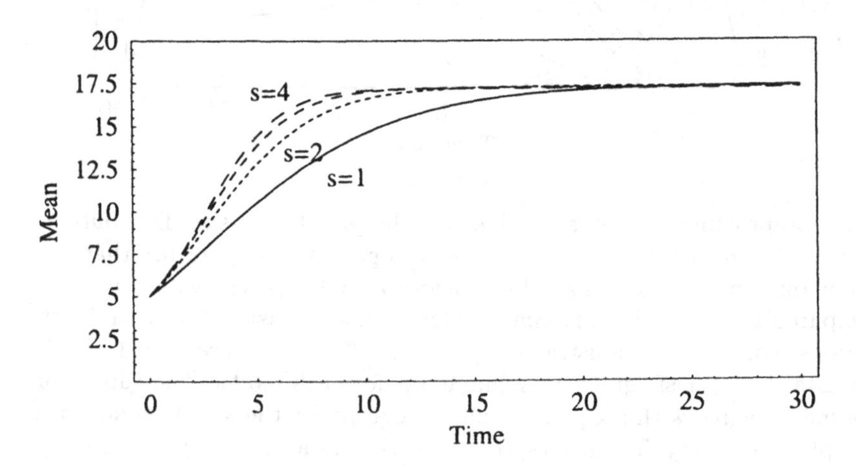

Figure 6.8. (Continued.)

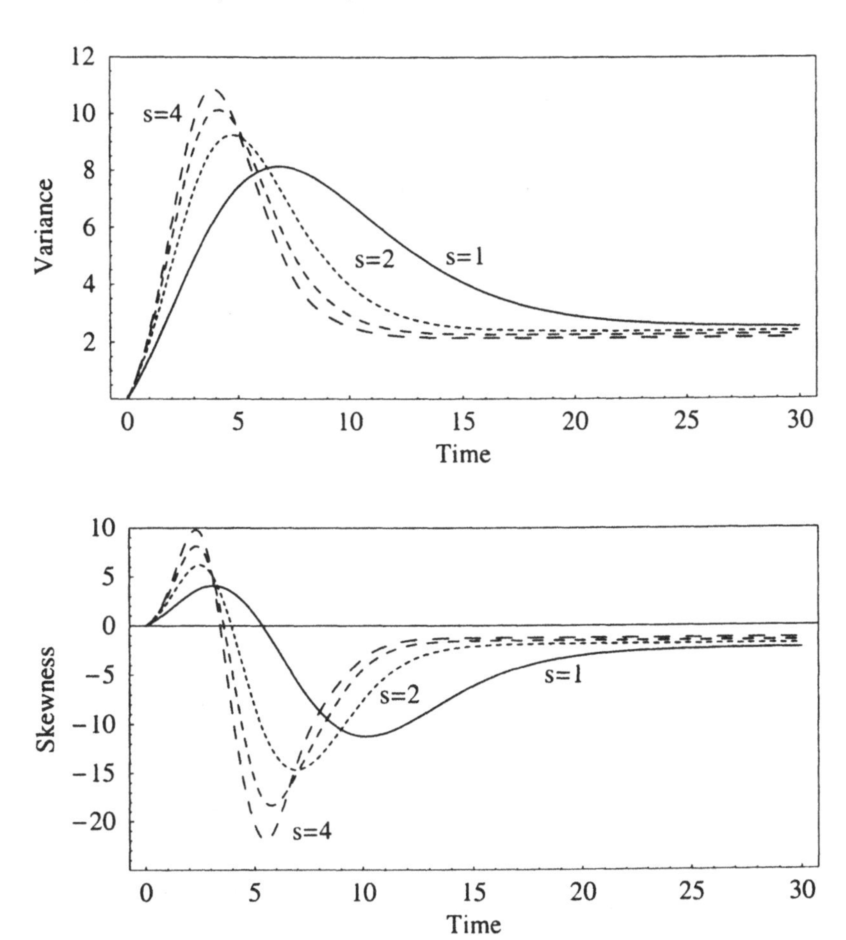

previously investigated in Chapter 5. The deterministic model in (6.5) with various fixed, small integer $s \geq 1$ was fitted to each data set using *Scientist* [67]. The best fitting curves for the three provinces has s values ranging from 2 to 5, however the models with $s = 2$ had the smallest average value of the mean squared error (MSE), a measure of goodness-of-fit. Figure 6.9 illustrates the fitted curves with $s = 1$ and $s = 2$ for overijl and dr. Research is in progress in fitting the mean value functions of the stochastic model to data.

Table 6.2 lists the parameter estimates, with their standard errors, and the mean squared errors for the models with $s = 1$ and $s = 2$ for each of the three provinces. For each province, the model with $s = 2$ had a better fit, i.e. lower MSE, a lower carrying capacity, K, and a lower net intrinsic rate, a.

Figure 6.9. Data and fitted logistic models with $s = 1$ and $s = 2$ for muskrat catch in two adjacent provinces, Overijssel and Drente.

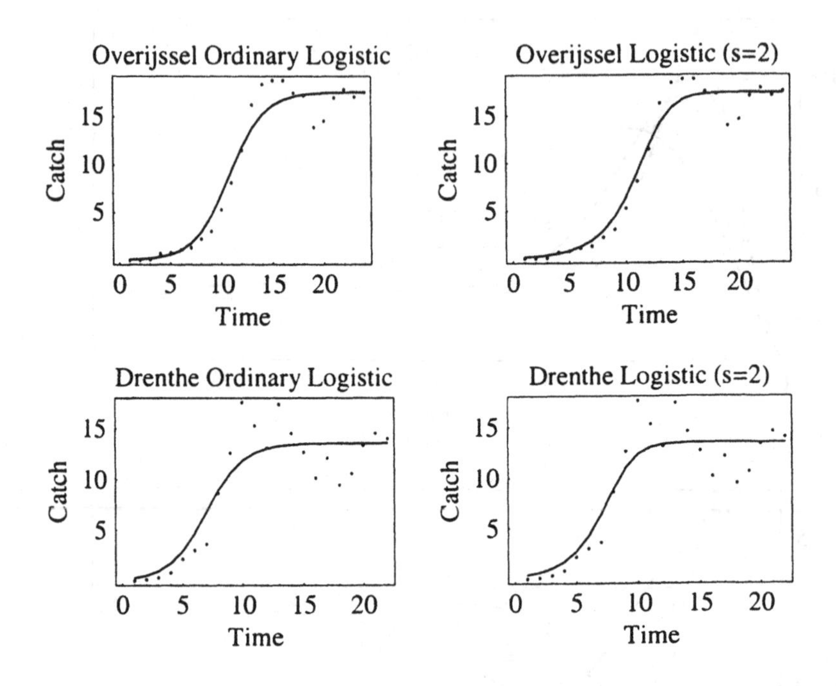

Table 6.2. Parameter estimates, with standard errors, and mean squared errors from fitting NBD models with $s = 1$ and $s = 2$ to the muskrat catch data for three adjacent eastern provinces.

Province	K	se	a	se	b	$MSE \times 10^{-6}$
$s = 1$:						
gelderl	48669	4104	0.408	0.125	8.38×10^{-5}	72.7
overijl	17595	601	0.583	0.102	3.31×10^{-5}	2.7
drente	13590	796	0.644	0.225	4.74×10^{-5}	6.2
for $s = 2$:						
gelderl	46216	2959	0.334	0.076	1.56×10^{-10}	64.3
overijl	17388	483	0.429	0.066	1.42×10^{-9}	2.1
drente	13503	695	0.483	0.163	265×10^{-9}	5.4

For comparative purposes, we assume as before in Section 5.7.2, a density independent death rate function with parameters $a_2 = 0.25$ and $b_2 = 0$ for each province. This implies that the birth rate function is density dependent with resulting parameters $a_1 = a + 0.25$ and $b_1 = b$. The assumed death

Figure 6.10. Estimated birth rate functions for three stochastic models; LBID, NBD for $s = 1$, and NBD for $s = 2$; and the assumed common death rate function for muskrat catch in two adjacent provinces.

A. Overijssel.

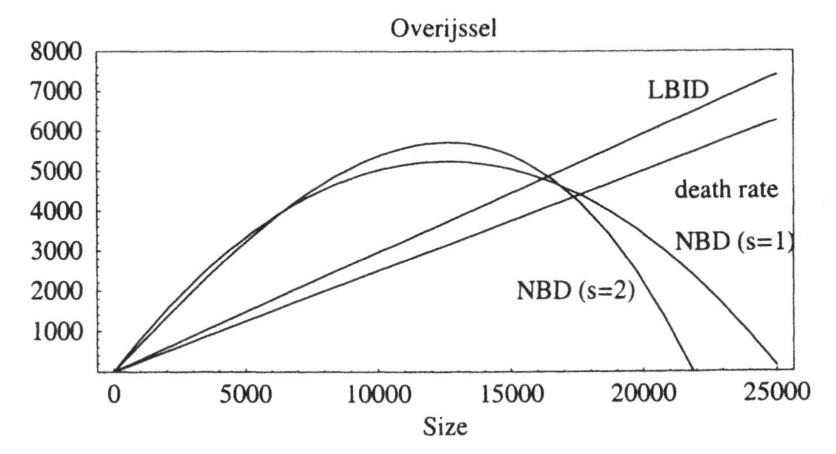

B. Drente.

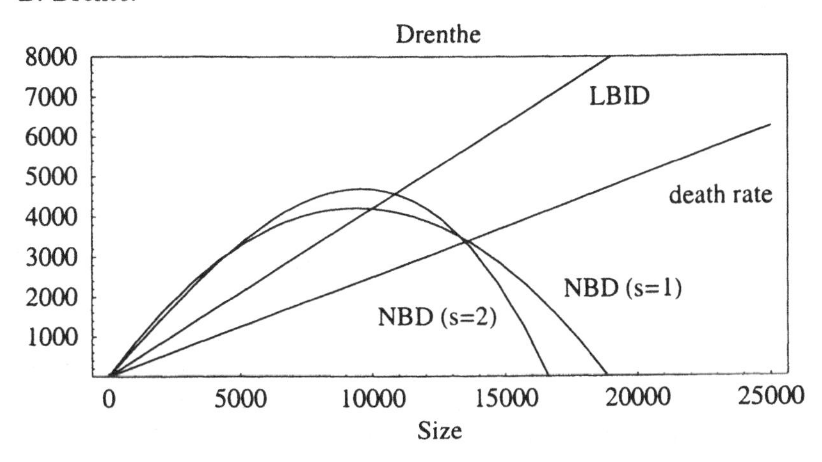

rate function and the estimated birth rate functions for three models, namely the LBID from Chapter 5 and the NBD with $s = 1$ and $s = 2$, are illustrated in Figure 6.10 for overijl and dr provinces.

These parameter estimates may be substituted into (6.17) and (6.18) to obtain approximate mean and variance functions. As an illustration, Figure 6.11 illustrates the approximate mean and variance functions for the model with $s = 1$ and $s = 2$ for overijl. The variance function is often a sensitive

Figure 6.11. Approximate mean and variance functions for the muskrat catch data in Overijssel under assumed NBD models with $s = 1$ and $s = 2$.

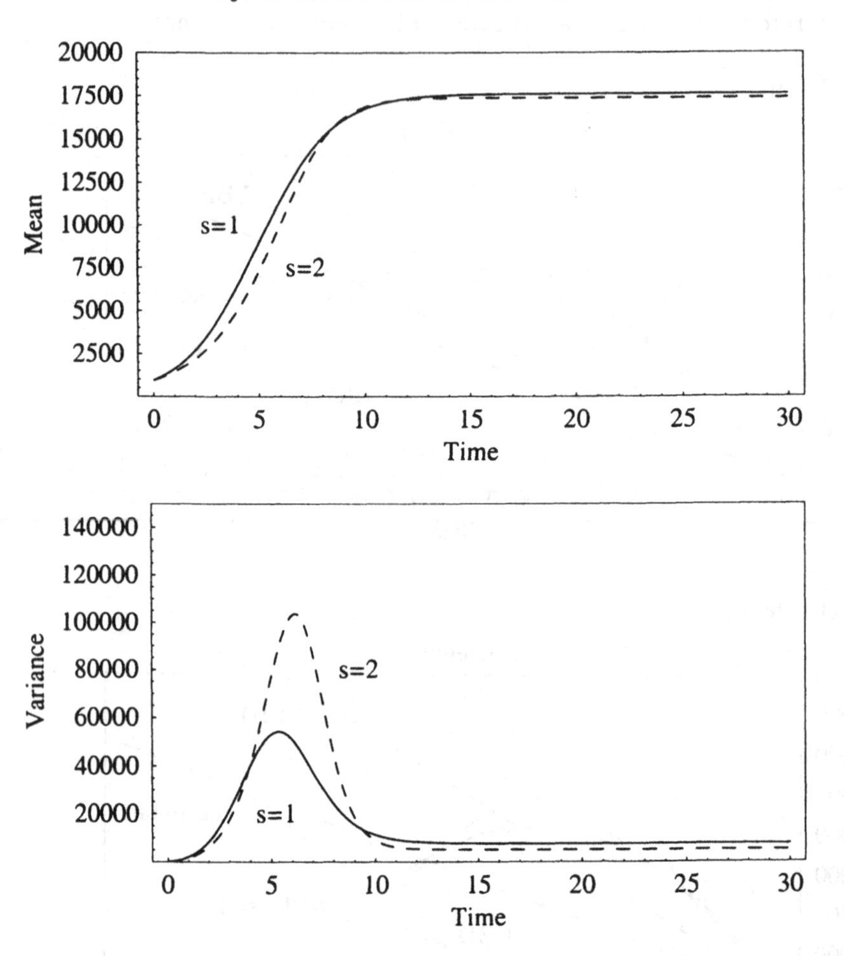

measure to use for model discrimination. Note in the present case that though the variances are quite large for some intermediate times, their equilibrium variances are too small to be consistent with the data in Figure 6.9. This suggests that models with additional sources of variability, for example so-called "environmental stochasticity" [14], might be considered for the variance function. Another striking feature in the data which is not explained by the model is the initial "overshoot" with a subsequent damped oscillatory approach to a quasi-equilibrium. Models that incorporate an overshoot phenomenon are under current investigation.

6.8 Appendices

6.8.1 Mean Time to Extinction

6.8.1.1 Direct Calculation

Consider now the time to extinction. Let T_x denote the time to extinction starting with population size $X(0) = x$, with density function $f_x(t)$ and cumulative distribution function $F_x(t)$. Using the duality between the population size and the waiting time variables, it follows that $T_x < t$ if and only if $X(t) = 0$. Therefore one has:

$$\begin{aligned} F_x(t) &= \text{Prob}[X(t) = 0 | X(0) = x] \\ &= p_0(t) \text{ given } X(0) = x. \end{aligned}$$

Taking derivatives, this implies that

$$f_x(t) = \dot{p}_0 \ (t)$$

Let E_x denote the mean of T_x. It follows by definition that

$$E_x = \int t \ \dot{p}_0 \ (t) dt,$$

and using (6.10) that

$$E_x = \mu_1 \int t p_1(t) dt. \tag{6.20}$$

This exact result for the mean time to extinction may often be evaluated numerically, and may be compared to some approximations [82].

6.8.1.2 Methods Based on Markov Processes

The mean time to extinction, E_x, found in (6.20) may also be obtained relatively easily using Markov process methodology. Let $\mathbf{R}^\dagger$ be a modified coefficient matrix of $\mathbf{R}$ in (6.11), obtained by deleting the first row and column. $\mathbf{R}^\dagger$ is thus the coefficient matrix of the Kolmogorov equations among the *transient* states. Let $\mathbf{M} = (m_{ij})$ be the matrix of so-called mean residence times. The element m_{ij} is defined as the total expected elapsed time that a population, which starts at size $X(0) = i$, will be of size j prior to the population becoming extinct. One can show [9, 60] that:

$$\mathbf{M} = -(\mathbf{R}^\dagger)^{-1} \tag{6.21}$$

The mean time to extinction given $X(0) = x$, denoted E_x, may be found by summing the m_{ij} elements, i.e.

$$E_x = \sum_j m_{xj}$$

In matrix notation, it follows that the column vector of E_x, denoted $\mathbf{M}_T$, is obtained by summing the rows of $\mathbf{M}$, i.e.

$$\mathbf{M}_T = \mathbf{M1} \tag{6.22}$$

where $\mathbf{1}$ is the vectors of 1's.

As an illustration, modifying the $\mathbf{R}$ coefficient matrix in our AHB example in Section 6.7.1, one has

$$\mathbf{R}^{\dagger} = \begin{bmatrix} -.306 & .285 & 0 & \dots & 0 \\ .044 & -.584 & .540 & \dots & 0 \\ \vdots & \vdots & \vdots & & \vdots \\ 0 & 0 & 0 & \dots & -.800 \end{bmatrix}$$

The solution for the $\mathbf{M}_T$ vector obtained from (6.21) and (6.22) is

$$\mathbf{M}_T' = 10^{12}[6.1267,\ 6.5781,\ 6.6149,\ 6.6182,\ 6.6186, \dots, 6.6186].$$

Note that $E_1 = 6.1267 \times 10^{12}$, as given previously in Section 6.7.1 using (6.10) and (6.20). The present approach is much simpler as it involves only simple matrix operations. E_x is an increasing function of x, but it does not increase measurably past $x = 5$.

6.8.1.3 Methods Based on Markov Chains

An equivalent result for $\mathbf{M}$ and $\mathbf{M}_T$ could be obtained using Markov chain methodology. Let a matrix $\mathbf{P}$ be defined such that

$$\mathbf{P} = \mathbf{I} + \mathbf{R}\Delta, \tag{6.23}$$

with $\mathbf{R}$ in (6.11) and $\Delta \le 1$ a scaling factor such that all the diagonal elements in $\mathbf{P}$ are positive. $\mathbf{P}$ is then the transition matrix for a discrete-time Markov chain. Deleting the first row and the first column, one has a matrix, denoted $\mathbf{Q}$, of transition probabilities among transient states. The elements of $\mathbf{Q}$ for $i \ge 1$ are

$$\begin{aligned} q_{i,i+1} &= \lambda_i \\ q_{i+1,i} &= \mu_i \\ q_{i,i} &= 1 - \mu_i - \lambda_i. \end{aligned} \tag{6.24}$$

The matrix $\mathbf{M}$, also called the fundamental matrix in Markov chain methodology, could be obtained as (see e.g. [9]):

$$\mathbf{M} = \Delta(\mathbf{I} - \mathbf{Q})^{-1} \tag{6.25}$$

As an illustration, with $\Delta = 0.1$, one has

$$\mathbf{Q} = \begin{bmatrix} .9694 & .0285 & \vdots & & 0 & 0 \\ .0044 & .9416 & \vdots & & 0 & 0 \\ \vdots & \vdots & & \vdots & \vdots & \\ 0 & 0 & \dots & & & .0285 \\ 0 & 0 & \dots & & .0800 & .9200 \end{bmatrix}.$$

from whence $\mathbf{M}$ and $\mathbf{M}_T$ may be calculated using (6.25) and (6.22).

6.8.2 *Quasi-Equilibrium Distribution*

6.8.2.1 Methods Based on Markov Processes

A quasi-equilibrium distribution was calculated previously based on the recurrence relationship in (6.13). There are several approaches involving direct matrix manipulation which simplify the numerical computations.

First, consider setting $\mu_1 = 0$ in the Kolmogorov equations in (6.10). This implies that $X = 1$ is a lower limit for population size, and thus it rules out extinction.

Let $\mathbf{R}^*$ denote the modified coefficient matrix, obtained by deleting the first row and column of $\mathbf{R}$ in (6.11) with the newly modified r_{11} element. Note that $\mathbf{R}^*$ differs from $\mathbf{R}^\dagger$ in (6.21) only in that the latter does not set $\mu_1 = 0$ in the r_{11} element. The equilibrium size vector, denoted

$$\boldsymbol{\pi} = (\pi_1, \pi_2, \dots, \pi_u),$$

is obtained by solving

$$\mathbf{0} = \boldsymbol{\pi}\mathbf{R}^*, \tag{6.26}$$

which follows directly from (3.17).

As an illustration, in our AHB example the $\mathbf{R}^*$ matrix in (6.26) is

$$\mathbf{R}^* = \begin{bmatrix} -.285 & .285 & 0 & \dots & 0 \\ .044 & -.584 & .540 & \dots & 0 \\ \vdots & \vdots & \vdots & & \vdots \\ 0 & 0 & 0 & \dots & -.800 \end{bmatrix}$$

and the equilibrium distribution is

$$\begin{aligned} \boldsymbol{\pi} = [&7.8E-12,\ 5.0E-11,\ 3.9E-10,\ 3.1E-9,\ 2.4E-8,\ 1.7E-7, \\ &1.2E-6,\ 7.1E-6,\ 3.9E-5,\ 1.9E-4,\ 8.5E-4,\ .0033,\ .0110, \\ &.0316,\ .0759,\ .1482,\ .2261,\ .2529,\ .1843,\ .0657]. \end{aligned}$$

The mean, variance, and skewness, denoted m, σ^2 and κ_3 respectively, for this population size distribution may be calculated directly from $\boldsymbol{\pi}$ to give

$$\begin{aligned} m &= 17.3564, \\ \sigma^2 &= 2.492, \text{ and} \\ \kappa_3 &= -2.210 \end{aligned} \tag{6.27}$$

which are consistent with the values observed in Figure 6.4.

6.8.2.2 Methods Based on Markov Chains

A second, equivalent approach based on Markov chain theory is to find first a modified $\mathbf{Q}$ matrix, say $\mathbf{Q}^*$, by setting $\mu_1 = 0$ in (6.24) to obtain $q_{11} = 1 - \lambda_1$. The matrix $\mathbf{Q}^*$ is then a transition matrix among the recurrent states, with equilibrium distribution obtained by solving (see e.g. [4, 9]):

$$\boldsymbol{\pi} = \boldsymbol{\pi}\mathbf{Q}^*. \tag{6.28}$$

For example, scaling again by $\Delta = 0.1$, which reduces the time step but does not affect the equilibrium distribution, the $\mathbf{Q}^*$ matrix is

$$\mathbf{Q}^* = \begin{bmatrix} .9715 & .0285 & \dots & 0 & 0 \\ .0044 & .9416 & \dots & 0 & 0 \\ \vdots & \vdots & & & \vdots \\ 0 & 0 & & & .0285 \\ 0 & 0 & & .0800 & .9200 \end{bmatrix}$$

The equilibrium distribution, $\boldsymbol{\pi}$, as previously given, could also have been obtained from (6.28).

6.8.2.3 Cumulant Approximations for the Quasi-Equilibrium Distribution

Bartlett *et al.* [7], in addition to utilizing (6.13) to find a quasi-equilibrium distribution, derive simple approximations for the moments of that distribution. Their results for the approximations to the mean, variance and skewness, are:

$$\hat{\sigma}^2 = a\gamma/2b^2, \tag{6.29}$$

where

$$\begin{aligned} \gamma &= b(a_1 + a_2)/a - (b_1 - b_2), \\ \hat{m} &= a/b - b\hat{\sigma}^2/a, \text{ and} \\ \hat{\kappa}_3 &= \hat{\sigma}^2(b_2 - b_1)/b, \end{aligned}$$

with a and b as defined in (6.4). These approximations have been widely used, and are illustrated also in [80, p. 35] and [82, p. 72–75]. As an illustration, these approximations were calculated for the parameters in Sections

6.2.2 and 6.7.1 which represent our standard AHB example. The percent errors in the approximations were obtained by comparing them to the exact results in (6.27). The results, with the percent error in parentheses, are

$$\hat{m} = 17.3661 \qquad (0.06)$$
$$\hat{\sigma}^2 = 2.344 \qquad (-5.90)$$
$$\hat{\kappa}_3 = -2.051 \qquad (7.20).$$

Cumulant approximations may also be obtained by setting the derivatives of the cumulant functions in (6.17) and (6.18) equal to 0, and then solving the resulting equations. Setting $\dot{\kappa}_1\ (t) = \dot{\kappa}_2\ (t) = 0$, and $\kappa_3 = 0$ for $s = 1$, yields the analytical solutions; from [44]:

$$_1\tilde{m} = (3a + \gamma_1)/4b \tag{6.30}$$
$$_1\tilde{\sigma}^2 = \left[a^2 + 4(a_2b_1 + a_1b_2) - a\gamma_1\right]/8b^2,$$

where

$$\gamma_1 = \left[a^2 - 8\left(a_2b_1 + a_1b_2\right)\right]^{1/2}$$

These approximations are usually more accurate than those in (6.29).

Consider a generalization by setting again $\dot{\kappa}_1\ (t) = \dot{\kappa}_2\ (t) = 0$, but using the skewness approximation $\kappa_3 = \kappa_2(b_2 - b_1)/b$ from (6.29). The analytical solutions are then, from [44]:

$$_2\tilde{m} = (3a + d + \gamma_2)/4b \tag{6.31}$$
$$_2\tilde{\sigma}_2 = \left[a_2 - d_2 + 4(a_2b_1 + a_1b_2) - (a + d)\gamma_2\right]/8b^2$$

where

$$\gamma_2 = \left[a^2 + d^2 - 2(a_1b_1 + a_2b_2) - 6(a_2b_1 + a_1b_2)\right]^{1/2}$$

with a, b, c, and d in (6.19). These approximations are almost always more accurate than those in (6.30) because of the inherent, though possibly only slight, skewness of the equilibrium distribution. As an illustration, the approximations for the standard AHB model, with percent errors in parentheses, are:

$$_1\tilde{m} = 17.3638\ (0.04) \qquad _2\tilde{m} = 17.3563\ (0.00)$$
$$_2\tilde{\sigma}^2 = 2.3624\ (-5.20) \qquad _2\tilde{\sigma}^2 = 2.4899\ (-0.07)$$

The analytical solutions in (6.29)–(6.31) for these cumulant approximations for the equilibrium size distribution for the ordinary logistic model, with $s = 1$, are obviously easy to use. We recommend the use of (6.31) because of the accuracy of its approximations. Simple, analytical solutions are not available for $s > 1$, however, approximations for the cumulants of such models may be obtained *numerically* by solving the set of equations in (6.17)–(6.19) with derivatives equal to 0, as also illustrated in [44].

7

Nonlinear Birth-Immigration-Death Models

7.1 Introduction

Consider now adding immigration at rate I to the nonlinear birth-death model. This new model is more realistic biologically in many applications, including the AHB and muskrat problems, however it is largely ignored in the literature. Perhaps there are two reasons for this. One is that the deterministic models become more unwieldy when I is included. The other reason is that inclusion of biologically reasonable levels of I seldom have a large impact on the qualitative *shape* of the deterministic solution. Consequently, in practice a statistical analysis would often fail to show that the inclusion of the parameter leads to a substantial improvement in the goodness-of-fit of the model to the data. In short, though there are often compelling biological reasons to include immigration, it is usually excluded from the model because it complicates the *deterministic* model and because its inclusion is typically not statistically "significant".

We will show, however, that the inclusion of immigration simplifies the analysis of the *stochastic* model. Moreover, the inclusion of immigration may have a striking impact on the stochastic solution, for example, on the probabililty distribution of population size. For these reasons, immigration is of considerable theoretical and applied interest in population modeling.

Consider again the density-dependent birth and death rate functions in Chapter 6, i.e.

$$\lambda_X = \begin{cases} a_1X - b_1X^{s+1} & \text{for } X < (a_1/b_1)^{1/s} \\ 0 & \text{otherwise} \end{cases} \tag{7.1}$$

$$\mu_X = a_2X + b_2X^{s+1} \tag{7.2}$$

for integer $s \geq 1$. Immigration at rate I, as defined in (3.2), is added to the model. This nonlinear model will be denoted NBID.

One immediate question is whether the immigration effect should also be density-dependent, e.g. at rate $I + a_3X - b_3X^{s+1}$. However the mathematical novelty for both the deterministic and the stochastic models comes only through inclusion of the constant I. Indeed, for $I = 0$, the proposed density-dependent immigration case with parameters a_3 and b_3 could be solved using Chapter 6, with a slight redefinition of the parameters a and b in (6.4). Hence, for present simplicity, we define immigration as the aggregate of all effects which tend to increase the population in a manner independent of its present size, i.e. which occur at some constant rate I independent of $X(t)$.

7.2 Deterministic Model

7.2.1 Solution for Deterministic Model

The deterministic model is clearly

$$\dot{X}(t) = I + aX - bX^{s+1} \tag{7.3}$$

with a and b defined as before in (6.4). The analytical solution for $s = 1$, is

$$X(t) = \left\{a + \beta\left[\left(1 - \delta e^{-\beta t}\right) / \left(1 + \delta e^{-\beta t}\right)\right]\right\} / 2b \tag{7.4}$$

where

$$\beta = (a^2 + 4bI)^{1/2}$$
$$\gamma = (2bX_0 - a)/\beta$$
$$\delta = (1 - \gamma)/(1 + \gamma)$$

The equilibrium value, or carrying capacity, for this case is:

$$K = (a + \beta)/2b \tag{7.5}$$

Equations (7.4) and (7.5) reduce readily to (6.5) and (6.6) in the special case of $I = 0$. The solution is given in a different but equivalent form in [4, p. 117]. This reference also discusses the case of $I < 0$, which instead of adding immigration to the model would incorporate "harvesting", frequently encountered in fisheries applications.

An analytical solution is available for $s = 2$, but is too cumbersome to be of present interest. Instead, for cases with $s > 1$, equation (7.3) will be solved numerically under the usual assumptions of positive I, a, and b. Obviously, as one compares such numerical solutions to the simple solution in (6.5), it is clear that the inclusion of constant immigration leads to substantial complications in the deterministic solution.

7.2.2 An Application to AHB Dynamics

We now consider comparing the NBID model to other models with the same carrying capacity. For example, we consider first a NBID model with $s = 1$

and the same intrinsic rates as the standard NBD model in Section 6.2.2, but with adjusted density-dependent coefficients, b_1 and b_2, to preserve the carrying capacity with immigration into the model. As a specific example, consider the NBID model with the specific parameter values

$$\begin{aligned} a_1 &= 0.30 \qquad a_2 = 0.02 \\ b_1 &= 0.012 \qquad b_2 = 0.004816 \\ I &= 0.25. \end{aligned} \tag{7.6}$$

These values hold the carrying capacity, from (7.5), at $K = 17.5$. The birthrate in (7.1) is 0 for all population sizes X exceeding $u = a_1/b_1 = 25$. These parameter values are closely related to those of an NBD model for AHB population dynamics studied in [56], which has the same a_i intrinsic rates, with $K = 17.5$ and $u = 25.6$.

The solution for the deterministic model, again with $X(0) = 2$, is

$$\begin{aligned} X(t) &= 8.3254 + 9.1749(1 - \delta e^{-\beta t})/(1 + \delta e^{-\beta t}) \\ &\text{where } \delta = 5.4364 \text{ and } \beta = 0.308571. \end{aligned} \tag{7.7}$$

This solution is plotted in Figure 7.1.

Figure 7.1. Deterministic solution, $X(t)$, and approximate mean value function, $\kappa_1(t)$, for the stochastic solution to the first assumed NBID model with initial size $X(0) = 2$.

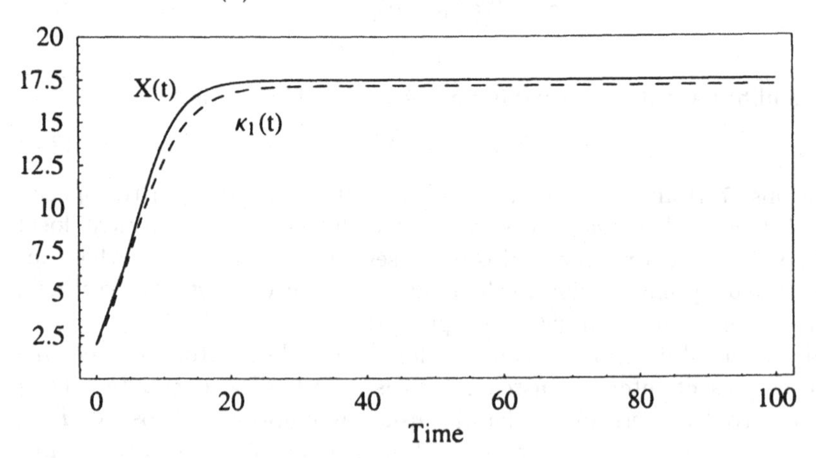

7.2.3 Application to Muskrat Population Dynamics

We illustrate also the NBID model with parameter values of I, a and b which are realistic for describing muskrat population dynamics. Unfortunately, precise estimates of the parameters are difficult to obtain by fitting (7.3) or (7.4) to data directly, due to statistical multicollinearity problems

[8]. The estimates hence are obtained by combining the previous results in Chapters 4 and 5 as follows.

When the LBID models are fitted to the initial population growth data, with relatively small population sizes, there is a strong statistical evidence of a significant immigration effect. The small population size enhances the statistical power for detecting immigration. The statistical significance of the immigration effect for three provinces of interest, namely gelderl, overijl, and drente, is apparent in the results of Section 5.7.2, (in fact the estimates of immigration exceed their corresponding standard errors by factors of roughly 2.5 to 5.0).

Consider now adding immigration to the NBD models fitted in Section 6.7.3. Though immigration is significant in the LBID model, the estimated immigration rates, of 342, 173, and 115, respectively, are tiny compared to the estimated carrying capacities, K, in Table 6.2 for the NBD model. In fact in each case the estimated yearly rate is less than 0.01 of its estimated K. Consequently, for these data the impact of immigration on the shape of the overall growth curve would be negligible. Therefore the NBID model illustrations in this chapter are based on the $\widehat{I}$ estimates from Section 5.7.2 and on the $\widehat{K}$ and $\widehat{a}$ estimates for the model with $s = 2$ from Section 6.7.3. These parameters are listed in Table 7.1.

Table 7.1. Parameter estimates for the muskrat population growth models.

A. Estimates with standard errors for the NBD model with $s = 2$.

Province	$\widehat{K}$	se	$\hat{a}$	se
overijssel	17,388	483	0.429	0.066
drenthe	13,503	694	0.483	0.163
gelderland	46,216	2,960	0.334	0.097

B. Estimates with standard errors for the LBID model with immigration.

Province	$\widehat{I}$	se	$\hat{a}$	se
overijssel	173	49	0.296	0.035
drenthe	115	46	0.421	0.055
gelderland	342	60	0.258	0.023

C. Assumed parameters for the logistic growth model ($s = 2$), NBID, with immigration, standardized to $K = 100$.

Province	I	a	a_1	a_2	$b(\times 10^5)$
overijssel	1.0	0.43	0.68	0.25	4.40
drenthe	0.9	0.48	0.73	0.25	4.89
gelderland	0.7	0.33	0.58	0.25	3.37

Two other modifications are made. Because the estimated carrying capacities differed so much, the carrying capacity was set for comparative purposes at $K = 100$ for each province. This envisions a subarea of the province of such size as to support a population of about 100 muskrats. The assumed value for I for the NBID model was obtained by scaling the $\widehat{I}$ estimate of the LBID to correspond to $K = 100$, i.e. $I = 100\widehat{I}/\widehat{K}$. The remaining parameter, b, is then obtained from the asymptotic population size in model (7.3), for which $\dot{X} = 0$ and $X(\infty) = K$. The solution for b from (7.3) follows as:

$$b = (I + aK)/K^{s+1}. \tag{7.8}$$

The solution for b is given in Table 7.1C for each of the provinces used to illustrate this NBID model.

Figure 7.2 illustrates the solution to model (7.3) with common initial value $X_0 = 5$ for the three provinces. These three provinces are contiguous and, as might be expected, have similar estimated deterministic growth curves after adjusting to the common K.

Figure 7.2. Deterministic NBID growth curves, standardized to carrying capacity $K = 100$, for three provinces, Overijssel, Drente and Gelderland.

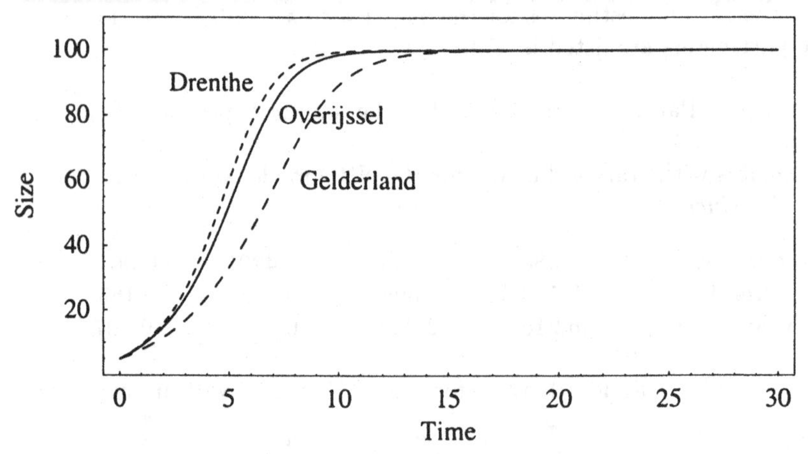

As previously with the muskrat data, we use the four-year life expectancy of a muskrat to assume a density-independent death rate with $a_2 = 0.25$/yr and $b_2 = 0$. The assumed birth rate from (6.4) is then $a_1 = a + 0.25$ with $b_1 = b$. These assumed birth- and deathrate functions are illustrated in Figure 7.3 for the three provinces. In the absence of immigration, the birth- and deathrate functions would intersect at size $X = 100$; with immigration the intersections are at X values slightly below 100.

Figure 7.3. Assumed common deathrate and individual birthrate functions for the stochastic NBID model for three provinces.

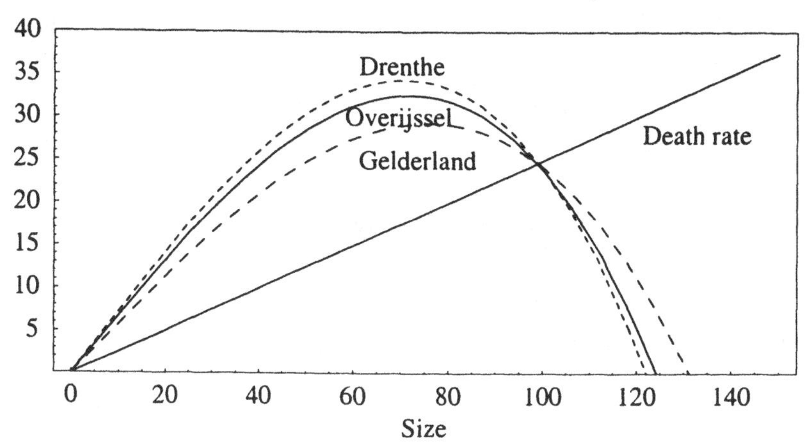

7.3 Probability Distribution for the Stochastic Model

7.3.1 Kolmogorov Equations

As before, the birthrate in (7.1) is 0 for $X > u$, where

$$u = (a_1/b_1)^{1/s}. \tag{7.9}$$

However, the population size may exceed u because of increases due to immigration, and hence the population size is unbounded, as with the LBID model in Chapter 5.

The Kolmogorov forward equations for the NBID model are

$$\dot{p}_x\,(t) = (\lambda_{x-1} + I)p_{x-1}(t) - (\mu_x + \lambda_x + I)p_x(t) + \mu_{x+1}p_{x+1}(t), \quad \text{for } x \geq 1, \tag{7.10}$$

with

$$\dot{p}_0\,(t) = \mu_1 p_1(t) - Ip_0(t).$$

The corresponding matrix $\mathbf{R}$ for the system of equations is infinite with elements, for $i,\ j \geq 0$:

$$r_{ij} = \begin{cases} r_{i,i+1} = \lambda_i + I & \\ r_{i,i-1} = \mu_i & \\ r_{i,i} = -(\lambda_i + \mu_i + I) & \\ r_{i,j} = 0 & \text{for } |i-j| > 1. \end{cases} \tag{7.11}$$

Approximate solutions for the $p_x(t)$ are again available, at least for initial periods of time, by truncating the distribution at some large upper limit U and solving the resulting finite system of equations.

7.3.2 Equilibrium Population Size Distribution

Following the development in [7], as outlined in Section 6.3.2, a population with immigration would satisfy the following recurrence relationship when it is in equilibrium:

$$\mu_x p_x(t) = (\lambda_{x-1} + I)p_{x-1}(t) \tag{7.12}$$

for all $x \geq 1$. An approximate distribution for the equilibrium probabilities, π_x, may be obtained numerically after truncating the distribution at some U. The procedure is illustrated subsequently.

7.4 Generating Functions

The "random variable" technique discussed in Section 3.6 which yields pde's for the generating functions is based on the assumption that the birthrate and deathrate functions, or more generally the rates of increase and decrease, are polynomials of the population size X. That necessary assumption is obviously not satisfied in the current migration model defined in (7.1), where the population size X may exceed the constant u in (7.9).

Consider therefore a modified model which changes the birthrate in (7.1) to

$$\lambda_X = a_1 X - b_1 X^{s+1} \tag{7.13}$$

for all $X > 0$. This modification introduces several issues. The birthrate for the deterministic model would be negative, which is not reasonable mechanistically, for $X > u$. This negative birthrate would have the effect of artificially reducing population growth due to immigration for large population sizes $X > u$.

The modification in (7.13) also creates basic theoretical problems for the stochastic model, our main concern at present. The underlying stochastic model formulation is based on infinitesimal probabilities of population size changes, as outlined in (3.2). Obviously, such probabilities could not be negative, as would be implied by using (7.13) for $X > u$. One solution to the theoretical problem is to introduce a constraint on immigration such that

$$I(x) = \begin{cases} I, & \text{for } x = 0, \ldots u, \\ 0 & \text{for } x > u. \end{cases} \tag{7.14}$$

This sets the upper bound of u on X.

In practice, most of our applications of interest of the original model in (7.1) yield a probability size distribution for which the probability of $X(t) > u$ is negligible for any t. In such cases the modification in (7.13) would have little practical effect on the distribution of $X(t)$. The effect of this modification will be illustrated subsequently with several examples.

With the birthrate in (7.13), deathrate in (7.2) and constant immigration, the intensity functions are

$$f_1 = a_1 x - b_1 x^{s+1} + I$$
$$f_{-1} = a_2 x + b_2 x^{s+1}$$

These may be substituted into the operator equations in (3.32) and (3.33) to obtain pde's for the generating functions. For example, for the case $s = 1$, the equations for the pgf and cgf are:

$$\frac{\partial P}{\partial t} = (s-1)(a_1 s - a_2)\frac{\partial P}{\partial s} + s(s-1)(b_1 s + b_2)\frac{\partial^2 P}{\partial s^2} + I(s-1)P \tag{7.15}$$

and

$$\frac{\partial K}{\partial t} = \left[\left(e^{\theta}-1\right)a_1 + \left(e^{-\theta}-1\right)a_2\right]\frac{\partial K}{\partial \theta} + \left[\left(e^{\theta}-1\right)(-b_1) + \left(e^{-\theta}-1\right)b_2\right]\left[\frac{\partial^2 K}{\partial \theta^2} + \left(\frac{\partial K}{\partial \theta}\right)^2\right] + I(e^{\theta}-1). \tag{7.16}$$

Each of these equations adds a term incorporating immigration to the corresponding equations (6.14) and (6.16) of the NBD model. Both generating functions remain analytically intractable.

Differential equations for the cumulants follow immediately because of the correspondence between (7.16) and (6.16). Indeed these equations for the NBD model are identical to those for the NBD model with the addition of the constant I on the right hand side. For example, the equations for $\kappa_1(t)$ and $\kappa_2(t)$ for the case $s = 1$ are:

$$\dot{\kappa}_1(t) = I + (a - b\kappa_1)\kappa_1 - b\kappa_2 \tag{7.17}$$

and

$$\dot{\kappa}_2(t) = I + (c - d\kappa_1)\kappa_1 + (2a - d - 4b\kappa_1)\kappa_2 - 2b\kappa_3 \tag{7.18}$$

with a, b, c and d as before in (6.19). Similarly the constant I would be added for any cumulant equation, $\dot{\kappa}_i(t)$, and for any power s.

7.5 Cumulant Functions

Approximate solutions to cumulant equations such as (7.17) and (7.18) may be obtained as described in Section 6.5 by setting higher order cumulants to 0. This adds another source of error to the approximations, in addition to the error from the modification in (7.13). The accuracy of the approximations is investigated in [45] by comparing these approximations to results obtained using the direct solutions to the Kolmogorov equations. As illustrated subsequently, this investigation suggests that the approximations are very accurate for low order ($i \leq 3$) cumulant functions in cases of practical interest, specifically when s is not large nor immigration massive.

7.6 Some Properties of the Stochastic Solution

One could argue that the *inclusion* of immigration changes the qualitative behavior of the stochastic model in a way that *simplifies* its analysis. Much of the simplification results from the fact that, although the population size could be 0 for periods of time, ultimate extinction is precluded because of immigration. Therefore, true equilibrium distributions exist, and there is no need to consider quasi-equilibrium distributions. Hence, the notion of conditional (on $X(t) > 0$) distributions of population size, as described in Section 6.6, is of little practical interest for the NBID model, and the issue of possibly conflicting results from the conditional and unconditional analyses is moot.

One practical difference in the transient size distributions with and without immigration is that the distributions with immigration tend to be more symmetric. This results from two qualitative effects. The first is that, with immigration, the probability of population size 0, i.e. $p_0(t)$, is not an increasing function. The increasing probability mass at $X(t) = 0$ for the NBD model without immigration tended to have a strong effect in increasing the skewness measure, $\kappa_3(t)$, which effect is absent in the present model. The second effect is that the (exact) population size distribution with immigration is not bounded above, often abruptly, by size u. This tends to give greater symmetry through a longer, more natural upper tail in the size distribution. The overall impact of these two effects is that population size distributions with immigration may often be adequately approximated by a Normal distribution, as illustrated subsequently. This may greatly simplify any subsequent statistical inferences concerning the model.

7.7 Illustrations

7.7.1 An Application to AHB Dynamics

Consider now the analogous stochastic model to the deterministic model illustrated in Section 7.2.2. The coefficient matrix **R** for this stochastic NBID model with parameters in (7.6) may be obtained from (7.1), (7.2) and (7.11) Approximate solutions for the transient distributions were obtained by truncating the system at $U = 40$. The numerical solutions for the $p_x(t)$ probability functions for all population sizes of practical interest are illustrated in Figure 7.4.

Figure 7.4. Approximate solutions for the $p_x(t)$ probability functions, $x = 1, \ldots, 22$; for the first NBID model of AHB population growth.

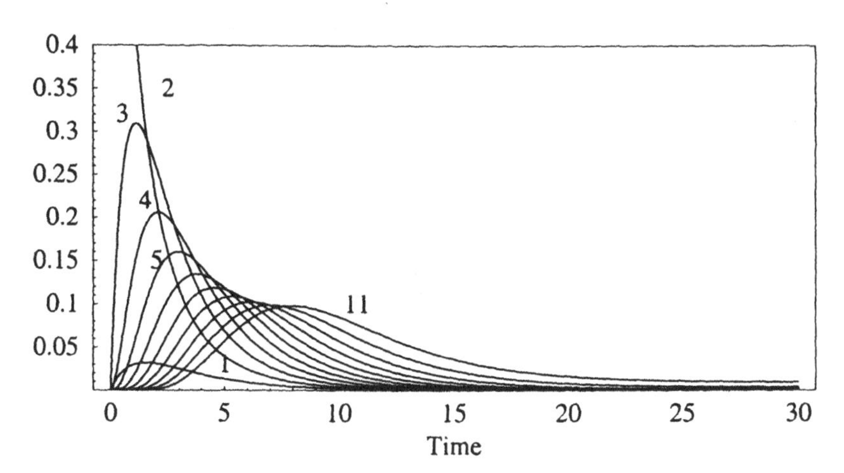

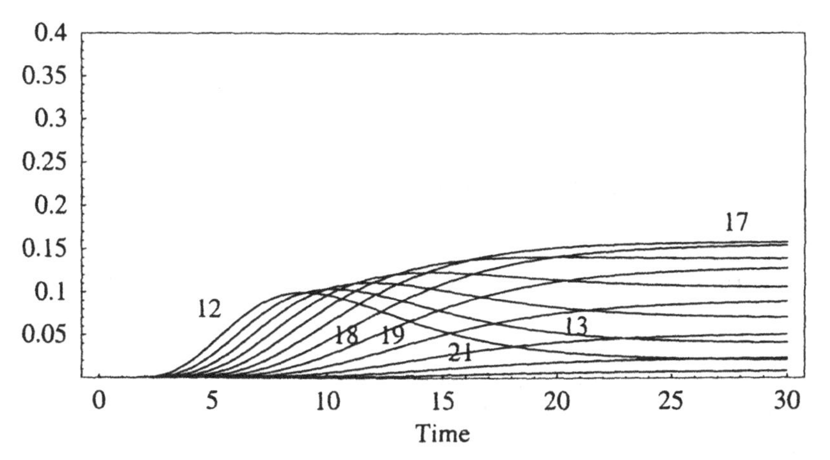

The equilibrium distribution, illustrated in Figure 7.5 together with the saddlepoint approximation from (3.11), is near-symmetric. A convenient measure of asymmetry is the index of skewness, γ, defined as

$$\gamma = \kappa_3/\kappa_2^{1.5}. \tag{7.19}$$

This measure for the present equilibrium distribution is $\gamma = -0.62$. As a contrast, the quasi-equilibrium distribution in Figure 6.3, which has the same intrinsic rates and carrying capacity, is $\gamma = -1.39$.

Cumulant functions obtained after modifying the birthrate to (7.13) are displayed in Figure 7.6. The sizes of the approximation errors due to this modification are obtained by comparing these solutions to those obtained directly from the transient probability functions in Figure 7.4, which are

Figure 7.5. Equilibrium population size distribution for the first assumed NBID model of AHB population growth, with a saddlepoint approximation.

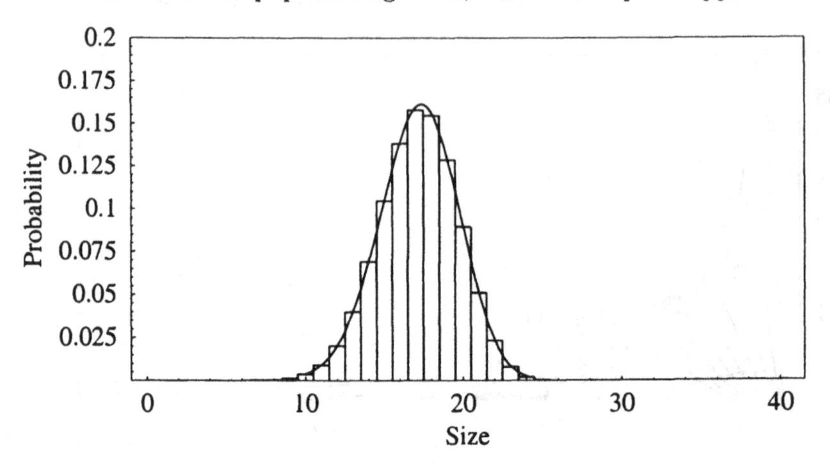

based on the birthrate in (7.1). These latter cumulant functions are also graphed in Figure 7.6, however they are indistinguishable from those with the modified birthrate for the mean and for the variance, and differ only slightly for the skewness functions. The effective equilibrium value (at $t = 100$) for the mean obtained from the Kolmogorov equations with (7.1) is $\kappa_1(100) = 17.1572$, and the cumulant approximation is the same to six significant places. The corresponding values for the equilibrium variance and skewness are $\kappa_2(100) = 6.1792$ and $\kappa_3(100) = -2.6657$, with the cumulant approximations having errors of only 0.01% and 0.6%, respectively.

Figure 7.6. Comparative approximate mean, variance and skewness functions for the first assumed NBID model, with standard (solid line) and modified (dashed line) birthrate functions.

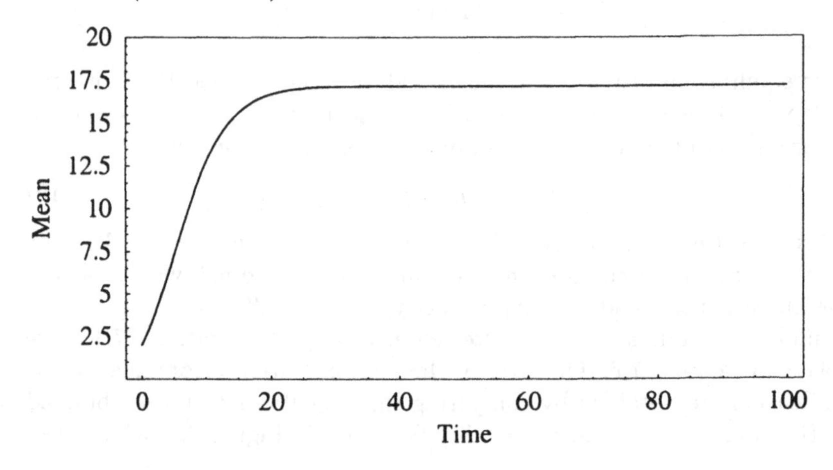

Figure 7.6. (Continued)

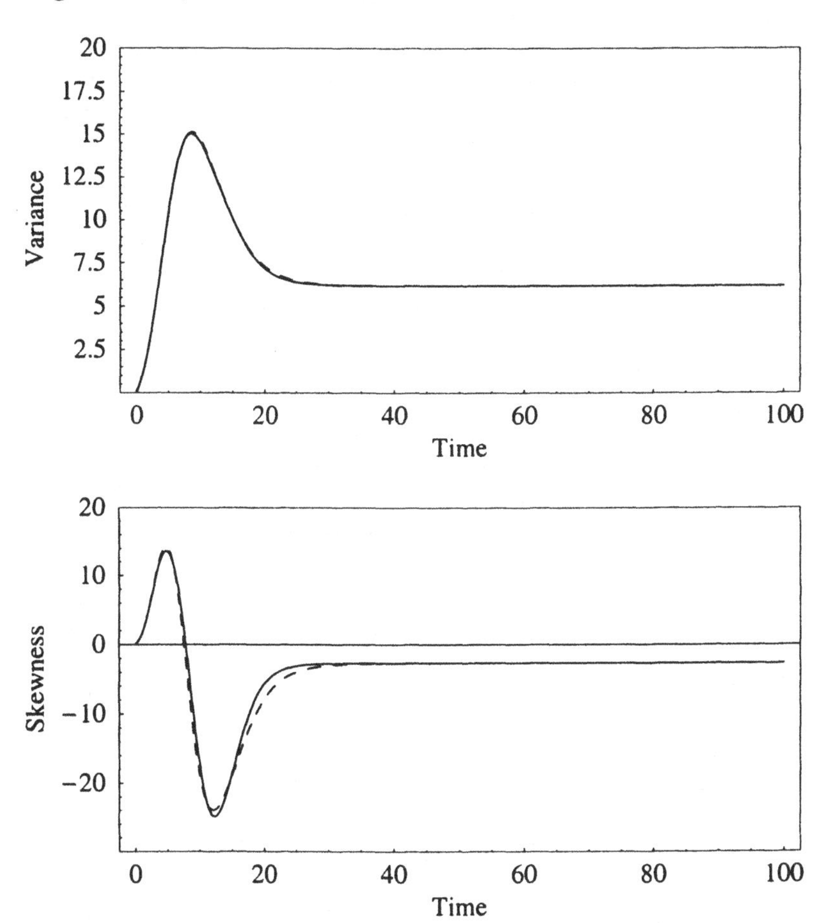

In summary, the errors introduced into the cumulant approximations are tiny for this example. This fact is not surprising in light of the solutions displayed in Figures 7.4 and 7.5, because there is negligible probability of a population size $X(t)$ exceeding $u = 25$.

7.7.2 A Second AHB Example

Consider a second NBID model, with parameter values

$$a_1 = 0.285 \qquad a_2 = 0.02$$
$$b_1 = 0.015 \qquad b_2 = 0.001$$
$$I = 0.2625.$$

These parameter values modify the standard NBD model in Section 6.2.2 in a different way, namely by the single change of reducing the intrinsic birthrate from $a_1 = 0.30$ to $a_1 = 0.285$ in order to compensate for the immigration effect of $I = 0.2625$. The carrying capacity for this new NBID model is again $K = 17.5$, but the birthrate is now 0 for all X exceeding $u = 19$.

The equilibrium size distribution obtained by solving the Kolmogorov equations truncated at $U = 40$ is given in Figure 7.7. The index of skewness for this distribution is $\gamma = -1.06$, which indicates a reduction in the asymmetry when compared to the analogous distribution without immigration in Figure 6.3 which has $\gamma = -1.39$. It is also apparent in Figure 7.7 that there is substantial probability of $X(t) > 19$ for large t, which in turn suggests that the approximation errors from the modification in (7.13) would be substantial. The equilibrium mean for the original process with (7.1) is 17.4789, and the modification reduces this mean to 17.3715, for an error of 0.6%. The corresponding values for the variances are 2.7157 and 2.3532, for an error of 13.3%, and for the skewnesses are -1.0301 and -2.0560, for an error of about 100%. Thus, the approximation error is large, as expected, for all cumulants other than the mean.

Figure 7.7. Equilibrium population size distribution for the second assumed NBID model.

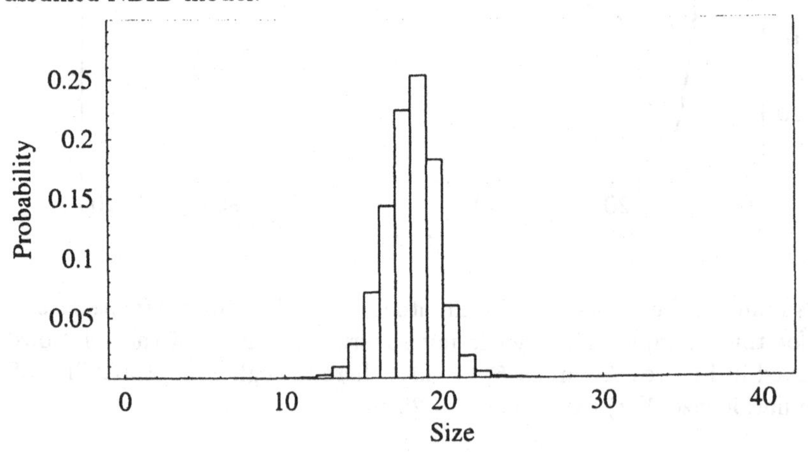

The two AHB examples indicate that there may be substantial errors in the cumulant approximations for relatively large levels of immigration. In practice, it may be useful to know *a priori* how accurate the cumulant approximations, which are easy to obtain, are without finding the exact values, which are more difficult to find. A simple monitoring test based on the Normal approximation is developed for this purpose in Section 7.8.1.3.

7.7.3 Application to Muskrat Population Dynamics

The equilibrium size distribution for the stochastic model was obtained from (7.12) for each of the three provinces, using parameter values in Table 7.1. The near-Normality of each distribution is remarkable, as illustrated for gelderl, a representative province, in Figure 7.8.

Figure 7.8. Equilibrium size distributions for the stochastic NBID model with the assumed parameter values for Gelderland province.

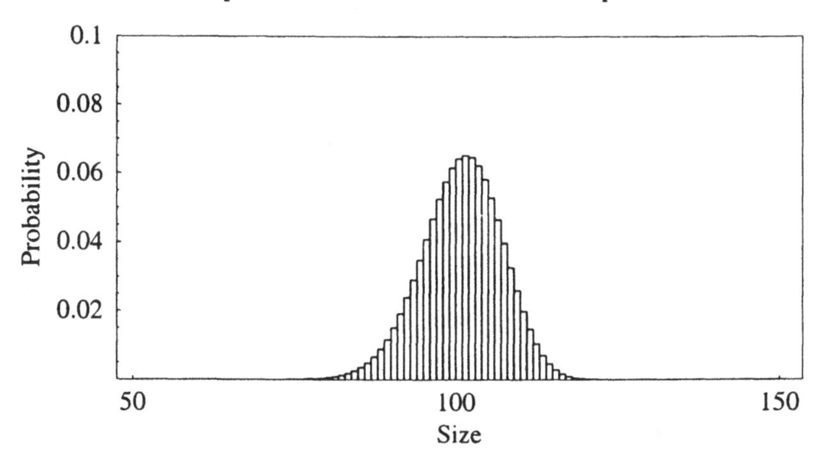

The approximate cumulant functions are portrayed for each of the provinces in Figure 7.9. The approximate mean value functions, $\widehat{\kappa}_1(t)$, are very accurate for each of the cases. The maximum relative error, denoted MRE, is calculated as

Figure 7.9. Approximate mean, variance and skewness functions for the stochastic NBID model for each of the three provinces.

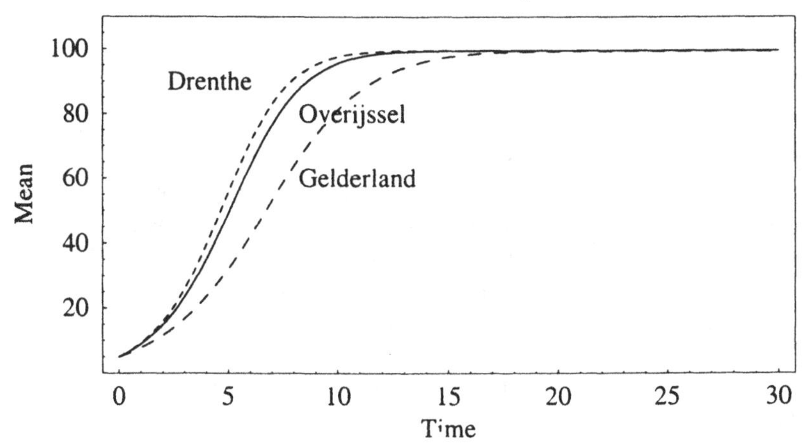

Figure 7.9. (Continued)

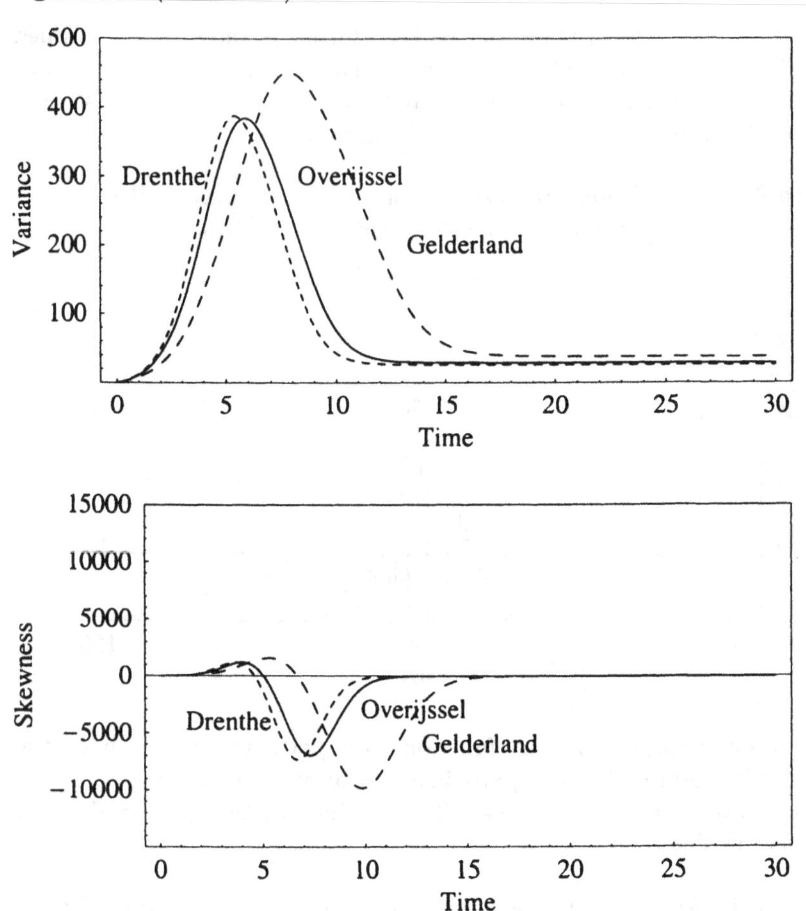

$$\text{MRE}_1 = \max\left\{100(\widehat{\kappa}_1(t) - \kappa_1(t))/\kappa_1(t)\right\}$$

with $\kappa_1(t)$ denoting the exact cumulant value from the Kolmogorov equations. For these data, MRE_1 is less than 0.15% for each province.

The approximate variance functions are also very accurate for most of the range of t for each case. The relative errors at the peak variances are all less than 1.0%. The maximum relative errors for the variances, MRE_2, rise to between 18 and 24% for the three cases. However, these large errors occur only for a small interval of time during which the absolute variances are relatively small. Figure 7.10 illustrates the exact, $\kappa_2(t)$, and the approximate, $\widehat{\kappa}_2(t)$, variance functions for the gelderland province which has the largest MRE_2, i.e. $\text{MRE}_2 = 24.0$. The excellent accuracy over most of the range is apparent even in this case.

Figure 7.10. Comparison of the exact, $\kappa_2(t)$, and approximate, $\widehat{\kappa}_2(t)$, variance functions for the NBID model with the assumed parameter values for Gelderland province.

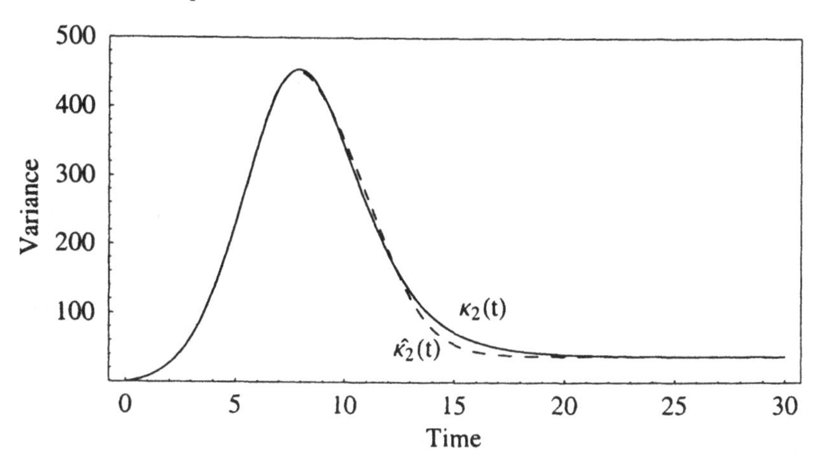

The maximum relative error for the skewness is not meaningful, as the function crosses 0. However, in each case the relative errors for the peak skewnesses are less than 16%.

In summary, the approximate cumulant functions are quite accurate for each of the three provinces with immigration of the size estimated for muskrat growth in the Netherlands.

7.8 Appendices

This section has two appendices. The first investigates the effect of various immigration levels in a NBID model. The second compares the LID, LBID, NBD and NBID models for a simple example. The two appendices develop different topics, and may be considered independently of one another.

7.8.1 Qualitative Effect of Immigration in the NBID Model

7.8.1.1 A Specific Family of Models

The behavior of a NBID model with various levels of immigration is studied in [56]. The parameters defining the family of interest are

$$a_1 = 0.75 \quad a_2 = 0.25 \quad \text{and} \quad K = 100 \tag{7.20}$$

which are suggested by the muskrat study. Models with three powers of s, namely $s = 1$, 2 and 3, were investigated. The model with $s = 1$ had the largest variances and skewnesses, and for convenience we describe here only the behavior of the $s = 1$ models.

Six levels of immigration, namely $I = 0.5, 1, 5, 10, 15$ and 20, were investigated. The level $I = 1$ best describes the estimated rates in the Netherlands for $K = 100$, with $I = 5$ as a practical upper limit. The higher immigration levels up to $I = 20$ were chosen to test the model under conditions of "massive" immigration. Reduced immigration is represented by $I = 0.5$.

The parameter for density dependency, b, was obtained for each of the six levels of immigration using (7.8) with the common parameters in (7.20). As in Section 7.2.3, we assume density-independent death, with $b_2 = 0$, hence the density-dependent birth rate has $b_1 = b$. The value of u, i.e. the population size for which the birthrate has decreased to 0, was calculated from (7.9) for each of the six cases. These parameter values are given for this family of models in Table 7.2.

Table 7.2. Numerical comparison of exact (κ_1, κ_2 and κ_3) and approximate ($\widehat{\kappa}_1$, $\widehat{\kappa}_2$ and $\widehat{\kappa}_3$) cumulants and index of skewness (γ_1) for equilibrium distributions for models with $s = 1$ and six levels of immigration, I.

I	b	u	κ_1	κ_2	κ_3	γ_1	$\widehat{\kappa}_1$	$\widehat{\kappa}_2$	$\widehat{\kappa}_3$
.5	.00505	148.5	99.50	49.76	-50.03	-.143	99.50	49.76	-50.02
1	.0051	147.1	99.52	48.78	-48.54	-.142	99.52	48.78	-48.54
5	.0055	136.4	99.61	42.17	-39.53	-.144	99.61	42.17	-39.53
10	.0060	125.0	99.69	36.07	-32.51	-.150	99.69	36.07	-32.51
15	.0065	115.4	99.74	31.56	-27.34	-.154	99.74	31.53	-27.88
20	.0070	107.0	99.99	30.68	5.71	.034	99.78	28.00	-24.57

Table 7.2 gives the first three exact cumulants κ_1, κ_2, and κ_3, for each of the six distributions with the different immigration levels.

The equilibrium mean, κ_1, increases as I increases, though it is bounded above by $K = 100$. However it is shown in [56] that for models with $s > 1$, κ_1 may exceed K in cases of large immigration.

The equilibrium variance, κ_2, decreases monotonically with I in Table 7.2, but again this is true only for the $s = 1$ model. The decrease is apparent visually in Figure 7.11. The skewness, κ_3, increases monotonically and in fact changes sign. The index of skewness, γ_1, is also given in Table 7.2 for each distribution.

7.8.1.2 Effect of Immigration on the Equilibrium Distributions

The exact equilibrium size distributions obtained from (7.12) for four immigration levels are represented in Figure 7.11.

Figure 7.11. Equilibrium population size distributions for the assumed stochastic NBID model with $s = 1$ and four levels of immigration, I.

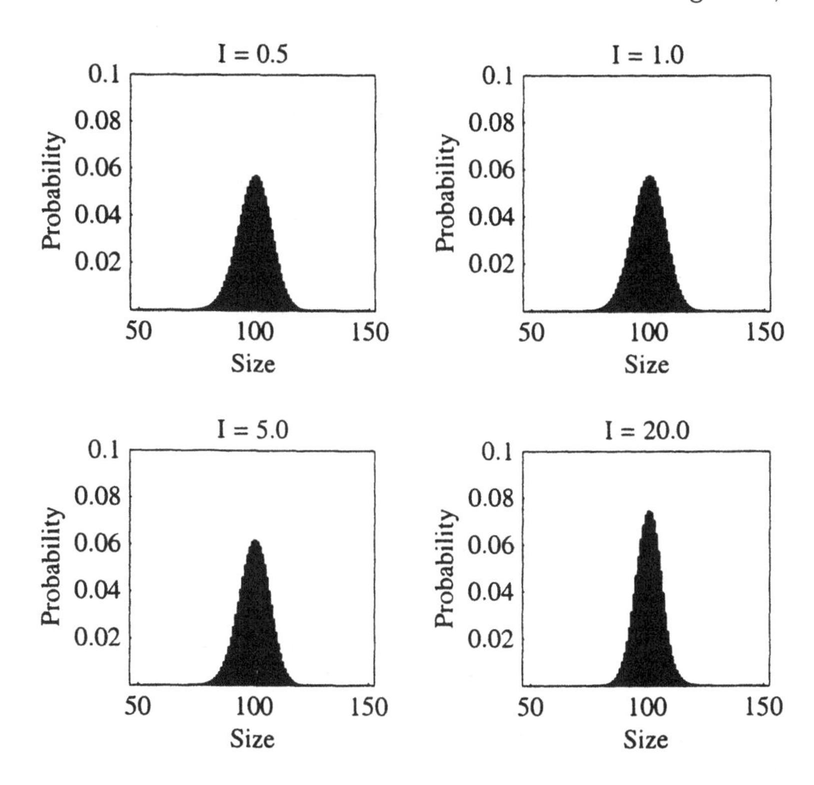

It is apparent in Figure 7.11 that all equilibrium distributions for $s = 1$ are approximately Normally distributed. For the $s = 2$ and 3 models, the index γ_1 is also small for any I. As a rule, therefore, the Normal distribution provides, for most practical purposes, an adequate approximation for the equilibrium distribution for any model with $s \leq 3$ and with any realistic level of immigration.

7.8.1.3 Effect of Immigration on Approximating the Equilibrium Distribution

Table 7.2 also gives the cumulant approximations, $\widehat{\kappa}_1$, $\widehat{\kappa}_2$, and $\widehat{\kappa}_3$, obtained as the solution to the truncated system of equations in Section 7.5 evaluated at $t = 100$. Table 7.3 gives the percent errors, $E_i = 100(\kappa_i - \widehat{\kappa}_i)/\kappa_i$, for these cumulant approximations for the six levels of I. For $I \leq 15$, the accuracy for the mean, variance and skewness approximations are less than 0.01%, 0.10%, and 2% respectively. The approximations seem inadequate only for massive immigration, in this case $I \geq 20$.

There is an apparent simple reason for the poor cumulant approximations with large I and, fortunately, an easy diagnostic test to ascertain when it would be a problem. As previously noted, the approximations have two sources of error, namely the modification of the birthrate in (7.13) and the truncation of higher order cumulants, as described in Section 7.5. The first source of error would have no measurable effect if the distribution of population size, X, has no meaningful support (i.e. probability) above u, as illustrated in Section 7.7.1. However one can show for the present parameter values that as I increases, with a corresponding decrease in u, there may be substantial probability that $X > u$.

The probability that $X > u$ in equilibrium may be estimated by exploiting the fact that the equilibrium distributions are approximately Normal. A standard Normal score could be calculated as

$$z = (u - \kappa_1)/\kappa_2^{1/2} \tag{7.21}$$

from whence the probability may be determined.

Consider, for example, setting a criterion that the probability of $X > u$ should be less than 1%. Standard Normal tables show that the upper 1%-ile is $z_{.01} = 2.33$. Therefore, for any model with $z > 2.33$ in (7.21), the first source of error, would have little practical effect.

Consider now the z values calculated from the cumulants in Table 7.2. All of the z values for models with $I \leq 15$ exceed 2.33, for example, the z value for $I = 15$ is $z = 2.79$. Specifically, the probability of a standard Normal variable exceeding 2.79, which in this case is an estimate that X exceeds $u = 115.4$, is only 0.003. This low probability is consistent with the visual observation in Figure 7.11 for this case.

Consequently, the cumulant approximations in Table 7.3 are all very accurate for immigration levels of $I \leq 15$. In fact, most of the error in the equilibrium approximation appears to be related to the first source, and the criterion $z > 2.33$ seems adequate to ensure accurate approximations. Of course, even more conservative criteria could be utilized based on reduced percentiles or on distribution-free bounds such as the Chebyshev bounds.

Table 7.3. Percent error of the cumulant approximations of the equilibrium distributions (E_1, E_2, E_3), maximum errors over time (M_1, M_2, M_3), and z^* criterion for model with $s = 1$ and six levels of immigration, I. (* denotes < .01, ws denotes wrong sign).

I	b	u	E_1	E_2	E_3	z^*	M_1	M_2	M_3
.5	.00505	148.5	*	*	0.02	6.95	0.06	9.80	9.04
1	.0051	147.1	*	*	0.01	6.81	0.02	3.04	5.10
5	.0055	136.4	*	*	*	5.67	*	0.04	0.42
10	.0060	125.0	*	*	*	4.21	*	*	0.20
15	.0065	115.4	*	0.10	1.99	2.79	*	0.10	0.18
20	.0070	107.0	0.21	8.74	ws	1.36			

Though the z score in (7.21) is a useful measure of the accuracy of the approximations, it is based on the exact cumulant values which in practice would not be known. In its stead, consider the estimate

$$z^* = (u - \widehat{\kappa}_1)/\widehat{\kappa}_2^{1/2}. \tag{7.22}$$

In our experience, one has $\widehat{\kappa}_2 < \kappa_2$, hence z^* would provide a conservative estimate of z and tend to underestimate the probability that $X > u$. Nevertheless, z^* is very close to z in the tail areas, e.g. with $z > 2.33$, and therefore z^* may be used instead of z in this range.

The value of z^* is listed for each model in Table 7.3. Note that it exceeds 2.33 handily for each case of $I \leq 15$. In practice, then, one could obtain the equilibrium approximations fairly easily from solving the differential equations, and then determine from (7.22) whether these approximations are sufficiently accurate for the desired purpose. This procedure also performs satisfactorily for this family of models with $s = 2$ and $s = 3$ [56].

In some applications, one might also consider the rough estimate

$$z^{**} = (u - K)/\sigma, \tag{7.23}$$

as a preliminary screen. The values of u and K may be easily calculated from (7.5) and (7.9), and perhaps σ could be guestimated from expert knowledge or prior experience. This would provide some indication of whether the approximations of equilibrium cumulants would be accurate even before the solution of the differential equations.

7.8.1.4 Effect of Immigration on Approximating the Cumulant Functions

Consider now the effect of immigration on the cumulants of the transient distributions of population size. For simplicity of practical application, we will base the comparisons on the approximate cumulant functions, $\widehat{\kappa}_1(t)$, $\widehat{\kappa}_2(t)$ and $\widehat{\kappa}_3(t)$.

Figure 7.12 illustrates these cumulant functions for the $s = 1$ family for the five immigration levels with $I \leq 15$, which were shown to give very accurate equilibrium approximations. As before, let $E_i(t)$ denote the percent error of the approximation of the ith cumulant at time t, and let M_i be defined for $i = 1$ and 2 as the maximum of $E_i(t)$ over $t > 0$, and for $i = 3$ as the larger of the two errors at the peak skewness values. The exact cumulant functions were calculated and the maximum errors are listed in Table 7.3 for the five cases.

Clearly the approximations are quite accurate in each case over the whole range of t. The largest error, for the variance with $I = 0.5$, is less than 10%. To put this into perspective, one can show that in each case the peak size of the variance is approximated to within 1%. The larger percent errors for the variance, in the order of 10% error, occur only for a narrow window of time

Figure 7.12. Approximate mean, variance and skewness functions for the assumed stochastic NBID model with $s = 1$ and five levels of immigration.

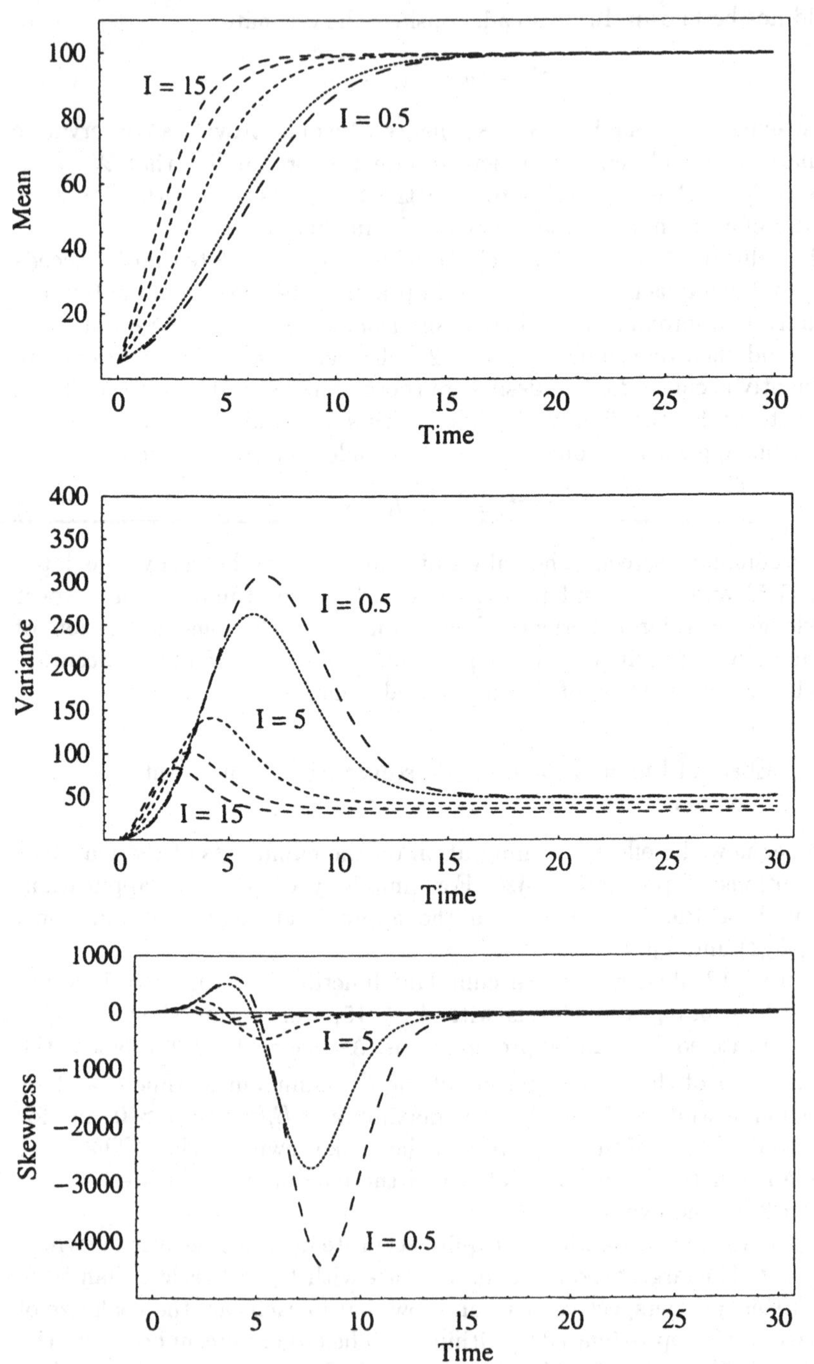

when the absolute size of the variance is less than 10% of its peak value. This characteristic was previously illustrated in Figure 7.10 for a set of closely related parameter values. Also, the maximum error in the skewness function is not large.

In brief, for those cases where the cumulant approximations are very accurate for equilibrium size (i.e. $z^* > 2.33$), the $\widehat{\kappa}_i(t)$ functions also approximate the extreme values very accurately and have substantial error only for a short range of time when the exact cumulants are small in absolute value. Hence, in practice, once one satisfies the z^* criterion for the equilibrium values, one may use the easily obtained cumulant approximations to investigate properties of the transient distributions.

7.8.1.5 Effect of Immigration on the Transient Distributions

Figure 7.12 gives the approximate cumulant functions for various levels of immigration in the $s = 1$ models. In each of these cases, the mean value functions rise more rapidly as I increases. The fact is not surprising, as these families were constructed by increasing b as I increases. Though the carrying capacity is fixed at $K = 100$, the larger value of I would have an immediate impact at any population size, whereas the larger value of b would have a substantial effect only for larger population sizes. As previously noted, the equilibrium mean, κ_1, also increases with I, hence the mean value, $\kappa_1(t)$, is an increasing function of I for any given $t > 0$.

The peak variance increases markedly as I decreases. The increase in the peak absolute value of the skewness as I decreases is even more pronounced.

In order to give a general feeling for the "worst-case" senarios, Figure 7.13 illustrates the population size distributions for four levels of I in the $s = 1$ models at their times of maximum variance. It is apparent in Figure 7.12 that, for each model, the skewness is at a near-maximum in absolute value at the time of maximum variance. As before, one can obtain an overall sense of the behavior of the transient distributions of population size, $X(t)$, for any of these cases by considering the extreme and the equilibrium distributions in Figures 7.11 and 7.13, combined with the cumulant functions in Figure 7.12.

It is clear in these figures that immigration has a strong effect in reducing the spread and the asymmetry of not only the equilibrium distribution but also of the transient distributions at any time t. The near-Normality of the transient distributions, even in the most extreme cases, seems remarkable.

7.8.2 Comparisons Among Single Population Growth Models

7.8.2.1 Introduction

There are substantial qualitative differences in the stochastic solutions of the four single population growth models, namely the LID, LBID, NBD and NBID models. In order to illustrate some of these differences, each of the four models were parameterized in the preceding chapters for the AHB illustration with initial value $X(0) = 2$ and with *deterministic* carrying capacity $X^* = K = 17.5$. Other parameters, particularly the birth rates, were chosen to give comparable mean value functions.

Figure 7.13. Distributions of population size for the assumed stochastic NBID model with $s = 1$ and four levels of immigration, I, at respective times of maximum variance.

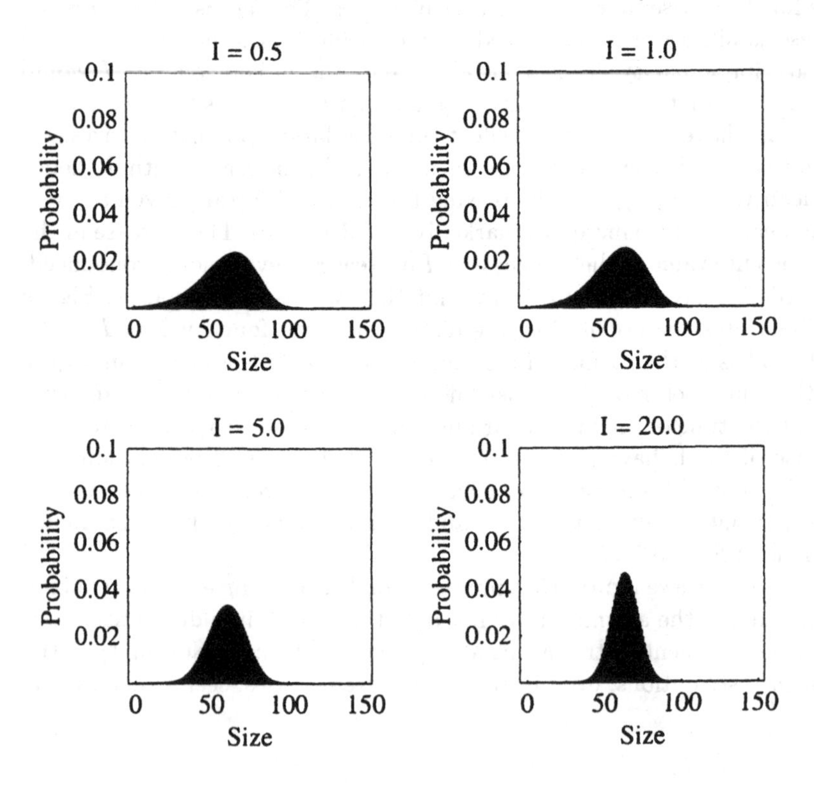

The *deterministic* solutions for the four models are illustrated together in Figure 7.14. As expected, the solutions for all but the two linear kinetic models (LID and LBID) differ in shape, for example the NBID has a more rapid initial growth spurt as a result of its larger initial derivative due to the addition of constant immigration. In general, however, the deterministic solutions for population size $X(t)$ for these four selected models have very similar qualitative shape characteristics over the range of t. Consequently, it may be difficult to discriminate among these models with very different kinetic assumptions on the basis of their deterministic solutions alone.

Suppose that these four deterministic models were actually fitted to a common data set. No doubt the differences in the four fitted solutions would be even smaller than those illustrated for the hypothesized models in Figure 7.14. It follows that discriminating between these four kinetic models on the basis of a single data set on population growth over time would be very difficult in practice. This raises the obvious question of how the *stochastic* solutions might differ.

Figure 7.14. Comparative deterministic solutions for four single population growth models, LID, LBID, NBD, and NBID, with common initial value, $X(0) = 2$, and carrying capacity, $K = 17.5$.

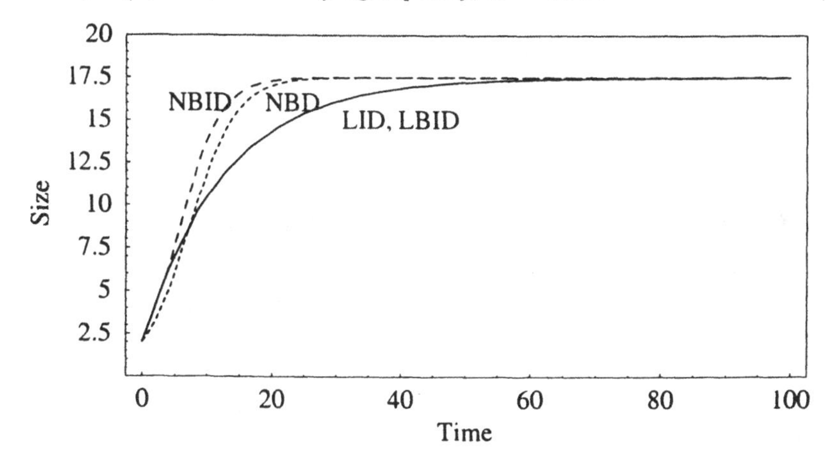

7.8.2.2 Comparison of Transient Probabilities Functions

Figure 7.15 gives the comparative solutions for $p_0(t)$ for the four kinetic models. Clearly, the solution for the NBD model is very different from the others. Under this model, size 0 (i.e. extinction) is an absorbing state, and hence $p_0(t)$ is an increasing function which slowly approaches 1. Because the other models include immigration, their asymptotic probabilities, $p_0(\infty)$, are some tiny positive values.

In general, there are major differences in the whole ensemble of probabilities, i.e. in the transient probability distributions, which may be useful in model discrimination. For convenience, first we examine only the cumulants of the transient distributions, and then compare the equilibrium distributions.

Figure 7.15. Comparative probability functions of population size 0, $p_0(t)$ for four stochastic, single population growth models.

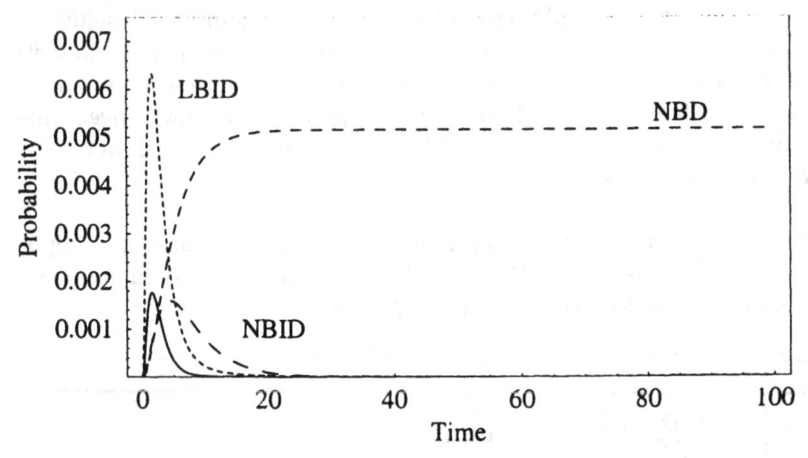

7.8.2.3 Comparison of Mean Value Functions

Figure 7.16 compares the mean value functions, $\kappa_1(t)$, for the four hypothesized models. The differences among these mean functions are even smaller than those observed previously among the deterministic solutions in Figure 7.14.

Figure 7.16. Comparative mean value functions, $\kappa_1(t)$, for four stochastic, single population growth models.

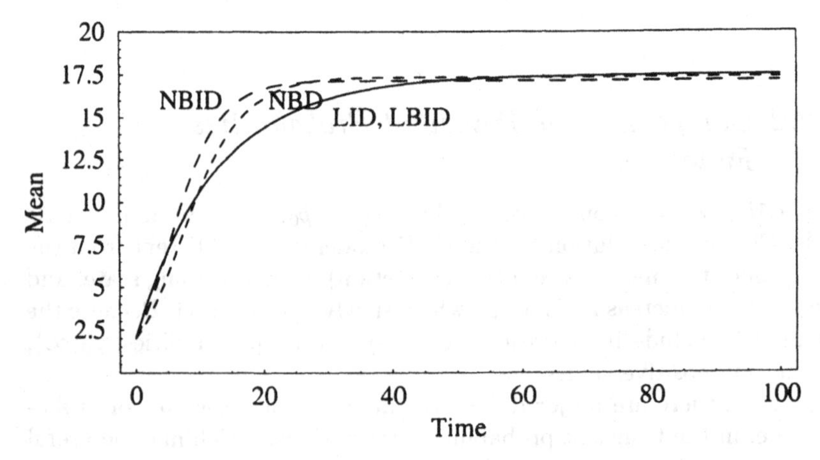

7.8.2.4 Comparison of Variance Functions

The four variance functions, $\kappa_2(t)$, are plotted in Figure 7.17, and they differ markedly. The two models with linear kinetics, LID and LBID, have variance functions which are monotonically increasing. With these two models, the model with births (LBID) has a substantially larger variances than the model without births (LID), as previously noted. The two models with the assumed nonlinear kinetics, NBD and NBID, have variance functions which rise initially to some peak, and then level off to much lower equilibrium values. The lower variances for large t are due to the "feedback" mechanisms of the density-dependent birth- and deathrate functions, which tend to stabilize the population sizes within certain equilibrium ranges.

Figure 7.17. Comparative variance functions, $\kappa_2(t)$, for four stochastic, single population growth models.

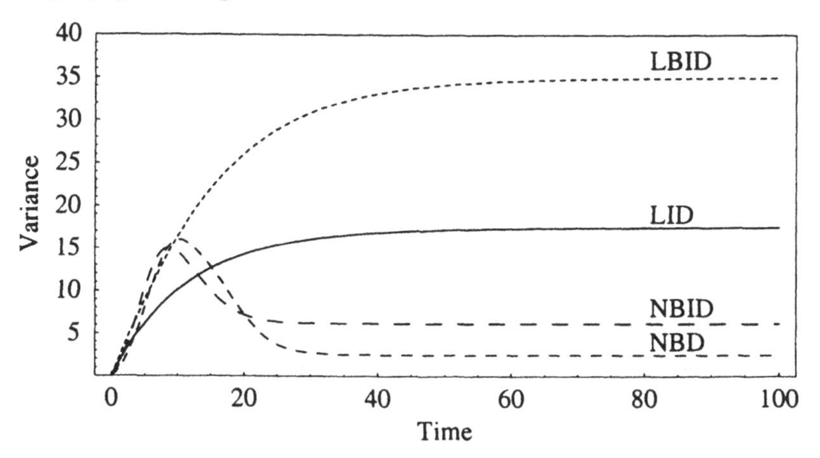

These striking differences might be useful in practice for model discrimination. It is standard practice in linear regression models to calculate the residuals, and then to examine them visually for various characteristic patterns. (see e.g. [70]). In principle, one could fit these kinetic models, specifically either $X(t)$ or $\kappa_1(t)$, to data and then examine the residuals. In particular, some pattern characteristic of either linear kinetics, with constantly increasing variance, or nonlinear kinetics, with a substantial reduction in the variance after some peak, might be apparent visually.

7.8.2.5 Comparison of Skewness Functions

Figure 7.18 illustrates the skewness functions, $\kappa_3(t)$, for the four models. The four functions differ by orders of magnitude for large t. The two nonlinear models have small equilibrium skewness measures, hence their equilibrium distributions are approximately Normal. The change in sign over time for the skewness functions of the nonlinear models is also noteworthy. Despite these pronounced differences, inferences based on the skewness measures are difficult to construct and are not very powerful statistically. Therefore these qualitative differences may not be useful in practice for model discrimination.

Figure 7.18. Comparative skewness functions, $\kappa_3(t)$, for four stochastic, single population growth models.

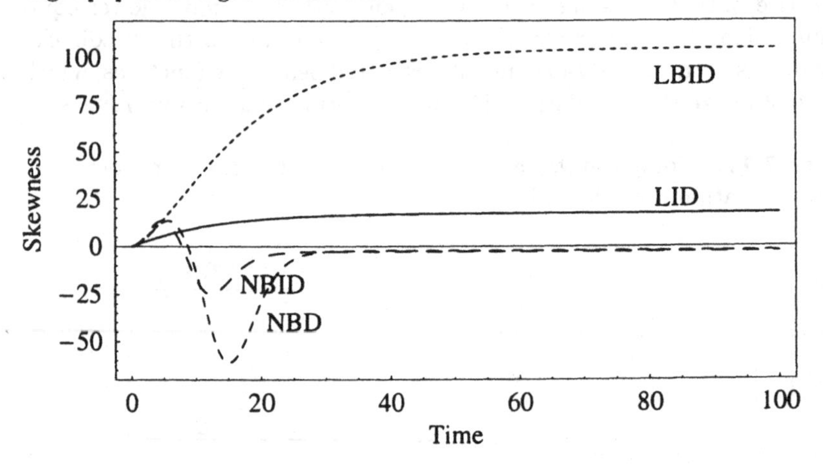

7.8.2.6 Comparison of Equilibrium Distributions

Figure 7.19 illustrates the equilibrium distribution for each of the four assumed models. It is apparent that the means of these distributions are very close, but that there are substantial differences in the equilibrium variances.

Consider discriminating among these four distributions on the basis of data that might be available after a process in question has reached for all practical purposes an equilibrium. One wellknown statistic for discriminating among such distributions is the variance to mean ratio, denoted r. A LID model generates a Poisson equilibrium distribution for which $r = 1$, a LBID model leads to a negative binomial equilibrium distribution for which $r > 1$, and the nonlinear kinetic model leads to reduced equilibrium variances for which $r < 1$. Hypothesis tests based on the sample variance to mean ratio have been developed to test for a Poisson distribution (i.e. $r = 1$). [105]. The negative binomial distribution and the distributions from the nonlinear kinetic models are immediate alternatives should the hypothesis of a Poisson be rejected.

Figure 7.19. Comparative equilibrium distributions for four stochastic, single population growth models.

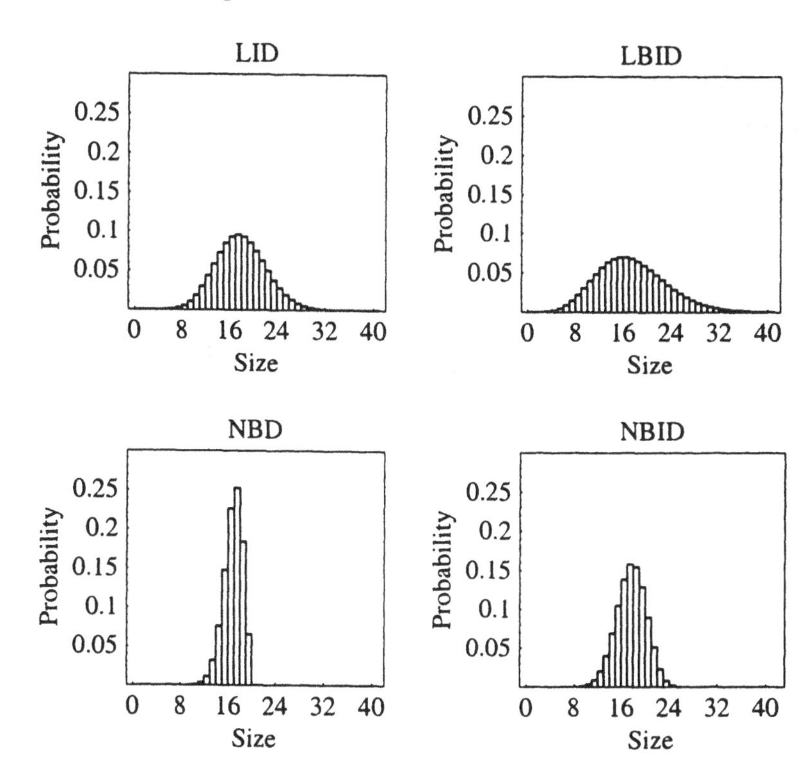

Part III Models for Multiple Populations

8

Standard Multiple Compartment Analysis with the Deterministic Model

8.1 Introduction

Compartmental modeling is widely used in pharmacokinetics and in physiological modeling, for which there is a vast literature. In addition to Jacquez [28], leading texts include [3, 22, 85]. As a simple example, the mercury bioaccumulation problem in Section 4.2.2 illustrates the need for a multicompartment analysis. The techniques have also been applied to ecosystem modeling (see e.g. [52]).

This chapter outlines the deterministic multiple compartment model, and illustrates its use on standard applications with two data sets. A matrix approach is used to facilitate model description, but only simple matrix operations are utilized. More advanced matrix analysis involving so-called eigenvalues and eigenvectors is referenced, but not incorporated. The two illustrations suggest the utility of a stochastic approach, the methodology of which is outlined in Chapter 9, with applications to linear kinetic systems following in Chapters 10–12 and to nonlinear systems in Chapters 13 and 14.

8.2 Deterministic Model Formulation and Solution

8.2.1 System of Differential Equations

The standard development of this model in a classical biomedical context often starts with the following definitions, all of them at time $t \geq 0$;

$f_{ij}(t)$ = flow rate of substance of interest to i from j (in units of mass/time) for $i, j = 0, \ldots, n$, with $i \neq j$; where 0 denotes the system exterior,

$k_{ij}(t) = f_{ij}(t)/X_j(t)$ = (proportional) "turnover rate" to i from j (in units of time^{-1}); for $i, j = 1, \ldots, n$;

$I_i(t) = f_{i0}(t)$ = flow rate (or immigration rate) to i from the exterior (in units of mass/time),

$\mu_j(t) = f_{0j}(t)/X_j(t)$ = turnover rate from j to the exterior (in units of time^{-1}). This is often denoted $k_{0j}(t)$.

In the pharmacokinetic literature [19], the order of the subscripts is reversed, i.e. $k_{ij}(t)$ denotes the flow rate from i to j.

For simplicity, we consider all of these rate functions to be constant, i.e. time invariant. The deterministic model assumes that the compartment sizes, $X_i(t)$, follow the system of (linear, first order) differential equations:

$$\dot{X}_1(t) = -(\mu_1 + k_{21} + \ldots + k_{n1})X_1 + k_{12}X_2 + \ldots + k_{1n}X_n + I_1 \tag{8.1}$$

$$\vdots$$

$$\dot{X}_n(t) = k_{n1}X_1 + k_{n2}X_2 + \ldots - (\mu_n + k_{1n} + \ldots + k_{n-1,n})X_n + I_n$$

For convenience, let

$$k_{jj} = -\left(\mu_j + \sum_{i=1, i\neq j}^{n} k_{ij}\right), \tag{8.2}$$

i.e. k_{jj} is the negative sum of all exiting rates from j. Then (8.1) reduces to

$$\dot{X}_1(t) = k_{11}X_1(t) + k_{12}X_2(t) + \ldots + k_{1n}X_n(t) + I_1$$

$$\vdots$$

$$\dot{X}_n(t) = k_{n1}X_1(t) + k_{n2}X_2(t) + \ldots + k_{nn}X_n(t) + I_n.$$

With the following matrix definitions:

$$\dot{\mathbf{X}}(t) = \begin{pmatrix} \dot{X}_1(t) \\ \vdots \\ \dot{X}_n(t) \end{pmatrix} \mathbf{X}(t) = \begin{pmatrix} X_1(t) \\ \vdots \\ X_n(t) \end{pmatrix} \mathbf{K} = \begin{pmatrix} k_{11} \ldots k_{1n} \\ \vdots \quad \vdots \\ k_{n1} \ldots k_{nn} \end{pmatrix} \mathbf{I} = \begin{pmatrix} I_1 \\ \vdots \\ I_n \end{pmatrix}, \tag{8.3}$$

the deterministic model may be written as

$$\dot{\mathbf{X}}(t) = \mathbf{KX}(t) + \mathbf{I}. \tag{8.4}$$

8.2.2 Solution to Deterministic Model

The formal solution to (8.4) may be expressed as

$$\mathbf{X}(t) = \exp(\mathbf{K}t)\mathbf{X}(0) + \int \exp[\mathbf{K}(t-s)]\mathbf{I}ds \tag{8.5}$$

where $\exp(\mathbf{K}t)$ is the matrix exponential defined as

$$\exp(\mathbf{K}t) = \mathbf{I} + \sum_{i=1}^{\infty} \mathbf{K}^i\mathbf{I}^i t^i / i! \tag{8.6}$$

Previously cited references may be consulted for a detailed matrix analysis of the model. Such analysis is omitted here, for present simplicity, however we note that under general conditions, the matrix $\mathbf{K}$ can be "diagonalized" and the solution $\mathbf{X}(t)$ in (8.5) may be written explicitly in terms of so-called eigenvalues and eigenvectors.

For subsequent purposes, we consider the solution for the $n = 2$ compartment model with coefficient matrix

$$\mathbf{K} = \begin{bmatrix} -(\mu_1 + k_{21}) & k_{12} \\ k_{21} & -(\mu_2 + k_{12}) \end{bmatrix}, \tag{8.7}$$

and hence with possible two-way flow, but without immigration, i.e. with $\mathbf{I} = \mathbf{0}$. The explicit solution for the matrix exponential, $\exp(\mathbf{K}t)$, for this model is

$$\exp(\mathbf{K}t) = \begin{bmatrix} \pi_{11}(t) & \pi_{12}(t) \\ \pi_{21}(t) & \pi_{22}(t) \end{bmatrix} \tag{8.8}$$

where

$$\begin{aligned} \pi_{11}(t) &= \left[(\lambda_1 + \mu_2 + k_{12})e^{\lambda_1 t} - (\lambda_2 + \mu_2 + k_{12})e^{\lambda_2 t}\right] / (\lambda_1 - \lambda_2) \quad (8.9) \\ \pi_{21}(t) &= k_{21}(e^{\lambda_1 t} - e^{\lambda_2 t})/(\lambda_1 - \lambda_2) \\ \pi_{12}(t) &= (\lambda_1 + \mu_2 + k_{12})(\lambda_2 + \mu_2 + k_{12})(e^{\lambda_2 t} - e^{\lambda_1 t})/k_{21}(\lambda_1 - \lambda_2) \\ \pi_{22}(t) &= \left[(\lambda_1 + \mu_2 + k_{12})e^{\lambda_2 t} - (\lambda_2 + \mu_2 + k_{12})e^{\lambda_1 t})\right] / (\lambda_1 - \lambda_2) \end{aligned}$$

with

$$\begin{aligned} \lambda_2, \lambda_1 = & -\Big\{(\mu_1 + k_{21} + \mu_2 + k_{12}) \\ & \pm \left[(\mu_1 + k_{21} - \mu_2 - k_{12})^2 + 4k_{12}k_{21}\right]^{1/2}\Big\}/2 \end{aligned} \tag{8.10}$$

The numbers λ_1 and λ_2 in (8.10) are the eigenvalues of the $\mathbf{K}$ matrix in (8.7).

A number of leading references in compartmental modeling, including [22, 28, 85], contain the explicit solutions to all 2-compartment and many 3-compartment models.

8.2.3 Example of Deterministic Model

As an example, consider a simple model with rates $\mu_1 = \mu_2 = 2$ and $k_{21} = k_{12} = 1$, previously used in a pharmacokinetic context in [47]. From (8.10) one has $\lambda_1 = -2$ and $\lambda_2 = -4$. The elements of the $\exp(\mathbf{K}t)$ matrix from (8.9) are

$$\begin{aligned} \pi_{11}(t) = \pi_{22}(t) &= (e^{-2t} + e^{-4t})/2 \quad \text{and} \\ \pi_{21}(t) = \pi_{12}(t) &= (e^{-2t} - e^{-4t})/2. \end{aligned} \tag{8.11}$$

Suppose one starts with an initial unit, e.g. a unit amount of tracer, in compartment 1, whereupon $X_1(0) = 1$ and $X_2(0) = 0$. It follows from (8.5), assuming $\mathbf{I} = 0$ that

$$X_1(t) = (e^{-2t} + e^{-4t})/2 \quad \text{and}$$
$$X_2(t) = (e^{-2t} - e^{-4t})/2.$$

If the bolus is introduced into compartment 2, with $X_1(0) = 0$ and $X_2(0) = 1$, it is clear using (8.5) that the solutions above would be reversed, as one would expect.

8.2.4 Solution for Equilibrium Values

One property of general interest that requires only basic matrix operations is the equilibrium solution $\mathbf{X}^*$. It follows by setting $\dot{\mathbf{X}}\ (t) = 0$ in (8.4) that

$$\mathbf{X}^* = -\mathbf{K}^{-1}\mathbf{I} \tag{8.12}$$

Equation (8.12) is the basis of a general ecological input-output analysis, called environ analysis [79]. In this analysis, the equilibrium amounts (or storages) in $\mathbf{X}^*$ for each compartment are partitioned linearly on the basis of the inputs, $\mathbf{I}$, into the compartments. The elements of $-\mathbf{K}^{-1}$ are interpreted as the multipliers which transform the inputs into equilibrium storages.

As a simple example, consider the previous two-compartment model. Substituting $\mathbf{K}$ in (8.7) into (8.12), the analytical solution for the equilibrium vector

$$\mathbf{X}^* = (X_1^*, X_2^*)'$$

is

$$X_1^* = \left[I_1(\mu_2 + k_{12}) + I_2 k_{12}\right]/d \text{ and} \tag{8.13}$$
$$X_2^* = \left[I_1 k_{21} + I_2(\mu_1 + k_{21})\right]/d,$$

where the determinant, d, is

$$d = (\mu_1 + k_{21})(\mu_2 + k_{12}) - k_{12}k_{21} \tag{8.14}$$

Hence for the previous model with $\mu_1 = \mu_2 = 2$ and $k_{21} = k_{21} = 1$, one has

$$X_1^* = 0.375 I_1 + 0.125 I_2 \text{ and} \tag{8.15}$$
$$X_2^* = 0.125 I_1 + 0.375 I_2.$$

Thus, according to this linear partitioning, every unit input into 1 (and also to 2 in this case) generates storages of 0.375 units in the recipient compartment and 0.125 in the other compartment. Such environ analysis is applied to a hydrology model in [79].

8.3 Illustrations

8.3.1 An Application to Bioaccumulation of Mercury in Fish

Consider again the mercury bioaccumulation data presented in Section 2.4 and previously analyzed in Section 4.2.2. Spacie and Hamelink [91] propose a model for bioaccumulation where heavy metal uptake and elimination occur through a tissue compartment, called compartment 1, where initial mercury processing takes place. This compartment is connected to other tissue with more permanent binding, conceptually a "storage" compartment, called compartment 2. The compartmental schematic for this conceptual model is given in Figure 8.1.

Figure 8.1. Schematic of LIDM model with storage compartment.

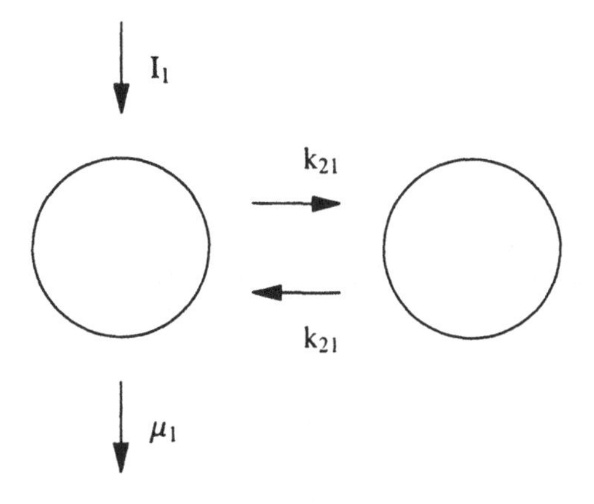

The present model has four parameters. All four parameters are estimable in principle in compartmental modeling [43]. However in the present application with only six data points per fish, estimation of all four parameters is impractical due to statistical multicollinearity [8]. Therefore, we make the simplifying assumption that $k_{12} = k_{21}$.

The set of deterministic equations for this storage model is then

$$\dot{X}_1\,(t) = -(k_{21} + \mu_1)X_1 + k_{21}X_2 + I_1 \tag{8.16}$$

$$\dot{X}_2\,(t) = k_{21}X_1 - k_{21}X_2,$$

with coefficient matrices of

$$\mathbf{K} = \begin{bmatrix} -(k_{21} + \mu_1) & k_{21} \\ k_{21} & -k_{21} \end{bmatrix} \quad \text{and} \quad \mathbf{I} = \begin{bmatrix} I_1 \\ 0 \end{bmatrix} \tag{8.17}$$

The solutions to these equations, assuming $X_1(0) = X_2(0) = 0$, is given in [28] as

$$X_1(t) = I_1 \left[(k_{21} + \lambda_1)(1 - e^{\lambda_1 t})/\lambda_1 + (k_{21} + \lambda_2)(1 - e^{\lambda_2 t})/\lambda_2 \right] / (\lambda_2 - \lambda_1)$$

$$X_2(t) = I_1 k_{21} \left[(1 - e^{\lambda_1 t})/\lambda_1 - (1 - e^{\lambda_2 t})/\lambda_2 \right] / (\lambda_2 - \lambda_1), \tag{8.18}$$

$$\text{where } \lambda_2, \lambda_1 = - \left[(2k_{21} + \mu_1) \pm \left(4k_{21}^2 + \mu_1^2 \right)^{1/2} \right] /2 \tag{8.19}$$

Clearly, (8.19) is a special case of the general solution for the eigenvalues in (8.10).

The concentration data, $c(t)$, for each (whole) fish was fitted to the aggregate model

$$c(t) = [X_1(t) + X_2(t)]/V \tag{8.20}$$

where V represents the mass of the fish. The estimable parameters in (8.20) are

$$I^* = I_1/V \tag{8.21}$$

for the proportional uptake into comp. 1, in units of μ gm Hg/gm dry wt/day, and the rate coefficients k_{21} and μ_1, in units of day^{-1}. The equilibrium concentration may be obtained by solving for $-\mathbf{K}^{-1}$ from (8.17) and using (8.12), or by using direct algebraic manipulation of (8.18) as

$$c(\infty) = 2I^*/\mu_1. \tag{8.22}$$

The fitted curves are illustrated in Figure 8.2. Each of them fits better, in terms of a lower residual mean square, than the corresponding fitted curves from the single compartment model in Section 4.2.2. The residuals, which are smaller under the new model, are also more randomly scattered. In particular the residuals for the concentrations at day 6, the last observed time, are tiny with no apparent bias under the new model. The least squares estimates for the proportional uptake rate of mercury, I^*, for the three fish are 0.46, 0.41 and 0.89. In each case, the estimate of uptake with the new model is somewhat larger than the corresponding estimate for the one-compartment model with the assumed "homogeneous" fish. The new estimates for the elimination rates, μ_1, from the three fish are 1.71, 0.88 and 1.73, respectively. These estimates are substantially larger than those for the previously assumed one-compartment model, and indicates a rapid turnover from the initial processing compartment (compartment 1) of the new model. The estimated equilibrium concentrations are 0.53, 0.93 and 1.02 μg Hg/gm dry wt, which are larger and more precise than the estimates from the previous model. In large part, this more precise estimate of the equilibrium concentration is due to the much smaller and apparently unbiased residuals for the terminal observations on day 6.

Figure 8.2. Observed and fitted values for the LIDM model of mercury bioaccumulation over time in three individual fish.

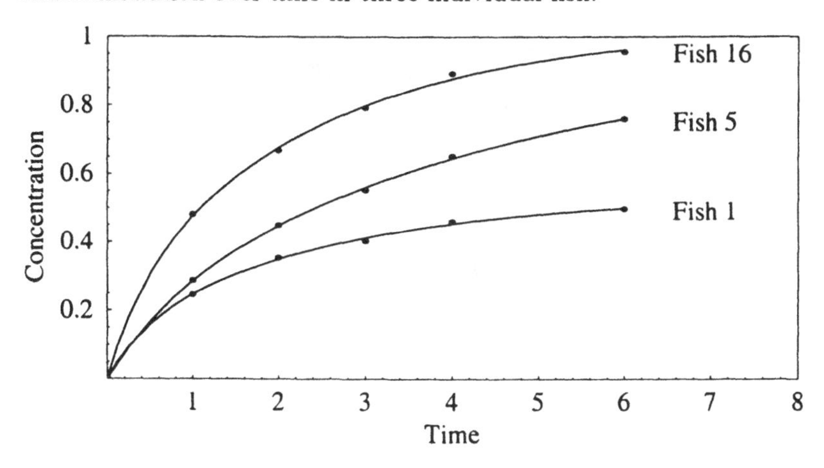

Clearly, other, more sophisticated models could be considered for mercury bioaccumulation, for example the model with $k_{21} \neq k_{12}$ would be a prime candidate. Nevertheless, even under the present limitation of only six data points per fish and with the resulting constrained model, there appears to be a strong case for a mercury storage compartment in these larger mosquito fish. Moreover, the rates of uptake and elimination may be determined with good precision for each fish and with reasonable consistency between fish. Further details are of the analysis are given in [57].

8.3.2 An Application to Human Calcium Kinetics.

As another illustration of a classical application of compartmental analysis, consider the calcium concentration-time curve described in Section 2.5 for an adult woman. A commonly used compartmental model for calcium kinetics is sketched in Figure 8.3., with compartment 1 representing the plasma compartment, compartment 2 the "soft" tissue, and compartment 3 the "hard tissue" including the bone. A calcium isotope was introduced to compartment 1 at $t = 0$, consequently we let $X_1(0) = X_1$ and $X_2(0) = X_3(0) = 0$, with $I = 0$. The deterministic model for the system is

$$\dot{X}_1(t) = -(\mu_1 + k_{21} + k_{31})X_1 + k_{12}X_2 + k_{13}X_3$$

$$\dot{X}_2(t) = k_{21}X_1 - k_{12}X_2$$

$$\dot{X}_3(t) = k_{31}X_1 - k_{13}X_3,$$

Figure 8.3. Schematic of the standard three-compartment LIDM model of calcium clearance.

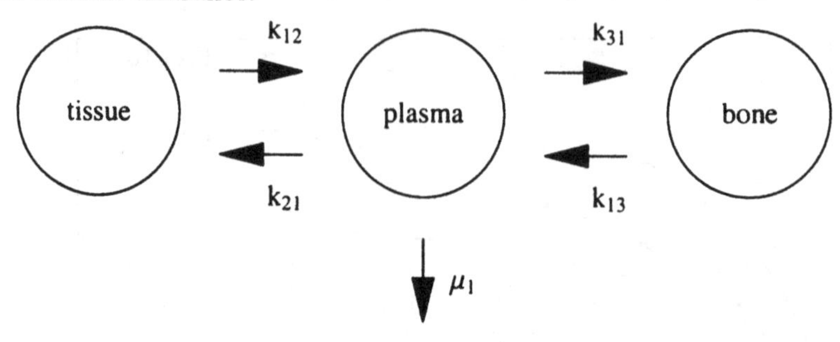

which yields the coefficient matrix

$$\mathbf{K} = \begin{bmatrix} -(\mu_1 + k_{21} + k_{31}) & k_{12} & k_{13} \\ k_{21} & -k_{12} & 0 \\ k_{31} & 0 & -k_{13} \end{bmatrix}. \tag{8.23}$$

The data in Section 2.5 represent the concentration, denoted $c_1(t)$, of labelled calcium in the plasma of this subject at various times, where

$$c_1(t) = X_1(t)/V_1,$$

with V_1 denoting the plasma volume. The parameters of the underlying model are the initial concentration, say $c_1(0) = c_0$, and the five rate coefficients. We used *Kinetica* [1] to fit the calcium model to these data, however any standard compartmental software package could be used.

It seems remarkable that one can estimate all six parameters of the model, with acceptable precision, from such data on a single compartment. The parameters, with the asymptotic standard errors in parentheses, are $\mu_1 = 0.0652$ (0.0022), $k_{21} = 2.611$ (0.251), $k_{12} = 2.998$ (0.289), $k_{31} = 0.389$ (0.036), $k_{13} = 0.162$ (0.014) and $c_0 = 693.9$ (17.0). The equation of the fitted curve, given by *Kinetica*, is:

$$\begin{aligned} c_1(t) = {} & 351.39 \exp(-5.835t) + 197.20 \exp(-0.375t) \\ & + 145.31 \exp(-0.0144t) \end{aligned} \tag{8.24}$$

which is illustrated in Figure 8.4. The estimated curve fits the data well (with a mean squared error of MSE=27.6). However, as indicated in [100], there is some indication of systematic, though small, deviations from the fitted curve.

Analogous stochastic models will be developed subsequently to describe the system kinetics. These stochastic models developed in Chapter 10 will add considerable insight into the underlying kinetic properties of the system, and also will generate alternative regression models which fit the data even better.

Figure 8.4. Observed data with fitted curve from multicompartment deterministic model of calcium clearance.

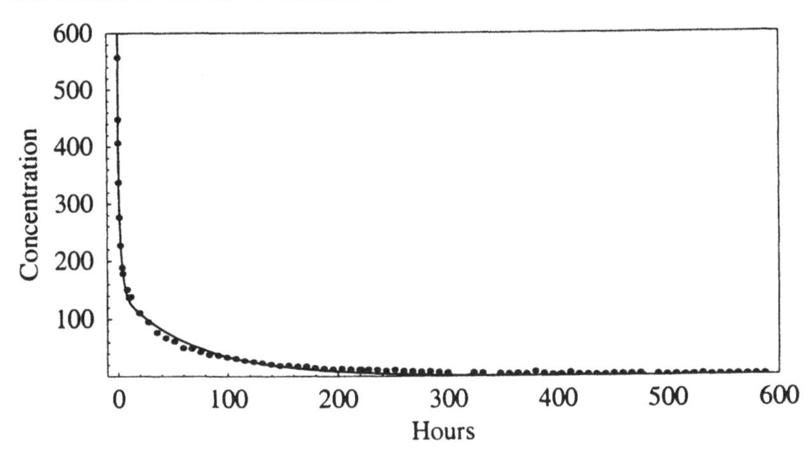

9

Basic Methodology for Multiple Population Stochastic Models

9.1 Introduction

This chapter generalizes the basic theoretical tools developed in Chapter 3 for single population stochastic models to the case of multiple population models. These tools also are considered extensively in the leading texts on applied stochastic processes, including [4, 9, 12, 82]. Jacquez [28] illustrates their use in a standard compartmental context.

9.2 Basic Assumptions

For present simplicity, we consider a model for only two populations linked by migration. The generalization to any arbitrary number of populations is straightforward but often tedious. Let

$X_i(t), i = 1, 2;$ denote the random size of the population in compartment i at elapsed time t,

$\mathbf{X}(t) = [X_1(t), X_2(t)]'$ be the random vector of sizes at t,

$$p_{x_1,x_2}(t) = \text{Prob}\,[X_1(t) = x_1, X_2(t) = x_2], \text{ i.e. the joint probability of sizes } x_1 \text{ and } x_2 \text{, respectively, at time } t, \tag{9.1}$$

$$\mathbf{p}(t) = [p_{00}(t), p_{10}(t), p_{20}(t), \ldots, p_{10}(t), p_{11}(t), \ldots], \text{ i.e. the bivariate probability distribution of } \mathbf{X}(t). \tag{9.2}$$

Our overall objective is to solve for the transient distribution $\mathbf{p}(t)$, at any $t > 0$, from simple assumptions concerning the dynamics of the two populations.

The standard birth-immigration-death-migration (BIDM) model has four types of possible changes over time in the population sizes. The first three types of changes are immediate generalizations of (3.2), with "instantaneous" and independent conditional probabilities of possible *unit* changes in small increments of time from t to $t + \Delta t$ as follow:

$$\begin{aligned} &1.\ \text{Prob}\ \{X_i \text{ will increase by 1 due to immigration}\} = I_i \Delta t, \\ &2.\ \text{Prob}\ \{X_i \text{ will increase by 1 due to birth}\} = \lambda_i X_i \Delta t, \\ &3.\ \text{Prob}\ \{X_i \text{ will decrease by 1 due to death}\} = \mu_i X_i \Delta t. \end{aligned} \tag{9.3}$$

The possible changes due to migration have conditional probabilities:

$$\begin{aligned} &4.\ \text{Prob}\ \{X_i \text{ will increase by 1 and } X_j \text{ will decrease by 1 due to} \\ &\qquad \text{migration}\} = k_{ij} X_j \Delta t, \text{ for } i \neq j. \end{aligned} \tag{9.4}$$

As before, we consider first a very basic model satisfying assumptions (9.3) and (9.4) in order to illustrate the methodology. Specifically, consider a model with immigration into compartment 1 only, at rate I_1, and with (linear) migration from 1 to 2 and (linear) death in compartment 2 at respective rates

$$k_{21} X_1 \quad \text{and} \quad \mu_2 X_2.$$

As before in Chapter 4, the absence of a birth mechanism simplifies the present analysis.

A schematic for this very restricted model is given in Figure 9.1. For simplicity, the initial population sizes are both assumed to be 0, i.e.

$$X_1(0) = X_2(0) = 0.$$

Figure 9.1. Schematic for simple immigration/migration model

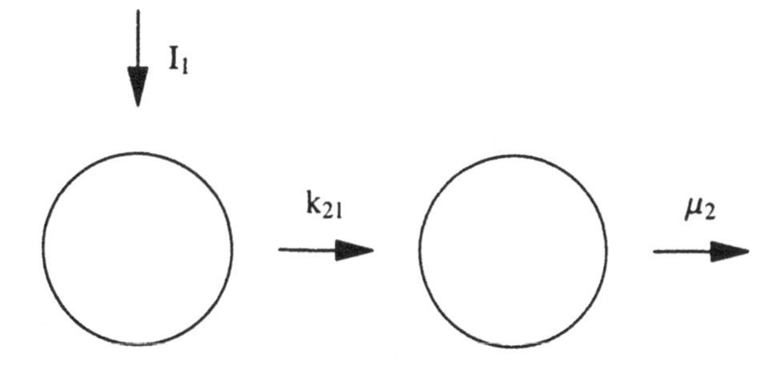

As an application of this special model, consider a house divided into two areas, as proposed in [40], and let $X_i(t)$, $i = 1, 2$, represent the number of roaches in area i at time t. Suppose that the house is fumigated at $t = 0$, say as a family leaves for vacation, hence the house is assumed at that time to be free of roaches, with $X_1(0) = X_2(0) = 0$. For further simplicity, suppose that roaches can enter the house only through area 1 (perhaps the kitchen) and that such immigration occurs at rate $I = 10$ roaches/wk. Also, suppose that migration occurs only from area 1 to area 2 (perhaps the living/dining area) at rate $k_{21} = 0.1$/wk, and that individual roaches "die" (i.e. depart) from area 2 only at rate $\mu_2 = 0.05$/wk.

Because the migration is uni-directional, area 1 in this problem is conceptually equivalent to the single compartment considered in Chapter 3. With the assumed rates, the "lifetimes" of roaches in the two compartments follow exponential distributions, with means of 10 and 20 wks, respectively. The population size, $X_1(t)$, was shown in Chapter 3 to follow a Poisson distribution, as indicated in (3.12) and (3.13). The primary questions of interest in this expanded, multi-population model relate to the (marginal) distribution of $X_2(t)$, and to its joint distribution with $X_1(t)$.

The system of equations defining the *deterministic* formulation described in Section 8.2.1 for this model is

$$\dot{X}_1(t) = I_1 - k_{21}X_1(t) \qquad (9.5)$$
$$\dot{X}_2(t) = k_{21}X_1(t) - \mu_2 X_2(t).$$

The solution, for $X_1(0) = X_2(0) = 0$, is

$$X_1(t) = (1 - e^{-k_{21}t})I_1/k_{21} \qquad (9.6)$$
$$X_2(t) = [(1 - e^{-\mu_2 t})/\mu_2 - (1 - e^{-k_{21}t})/k_{21}]I_1 k_{21}/(k_{21} - \mu_2).$$

With the assumed specific parameter values, the solution is

$$X_1(t) = 100(1 - e^{-0.1t}) \qquad (9.7)$$
$$X_2(t) = 200[2(1 - e^{-0.05t}) - (1 - e^{-0.1t})].$$

Some analogous stochastic solutions to the problem are developed subsequently in this chapter.

9.3 Joint Moments and Cumulants

The joint moments for the distribution of the random vector $\mathbf{X}(t)$ are defined as

$$\mu_{ij}(t) = \sum_{x_1,x_2} x_1^i x_2^j p_{x_1,x_2}(t). \qquad (9.8)$$

Joint cumulants, denoted $\kappa_{ij}(t)$, will be formally defined subsequently through the cumulant generating function. For present purposes, one can show that the first order cumulants, $\kappa_{10}(t)$ and $\kappa_{01}(t)$, give the two mean population size functions, and the second order cumulants, $\kappa_{20}(t)$, $\kappa_{11}(t)$, and $\kappa_{02}(t)$, represent the two variances and the covariance of the population sizes. Numerous relationships between low-order joint moments and cumulants are listed for multivariate distributions [5, 107]. These relationships for our bivariate case for up to second-order cumulants are:

$$\begin{aligned}
\kappa_{10}(t) &= \mu_{10}(t) \\
\kappa_{01}(t) &= \mu_{01}(t) \\
\kappa_{20}(t) &= \mu_{20}(t) - \mu_{10}^2(t) \\
\kappa_{02}(t) &= \mu_{02}(t) - \mu_{01}^2(t) \\
\kappa_{11}(t) &= \mu_{11}(t) - \mu_{10}(t)\mu_{01}(t),
\end{aligned} \tag{9.9}$$

all of which, except for $\kappa_{11}(t)$, are immediate generalizations of (3.10).

As an illustration, it will be shown later that $\mathbf{X}(t)$ for the roach migration problem has a bivariate Poisson distribution, with parameter functions

$$\begin{aligned}
\lambda_1(t) &= (1 - e^{-k_{21}t})I_1/k_{21} \\
\lambda_2(t) &= [(1 - e^{-\mu_2 t})/\mu_2 - (1 - e^{-k_{21}t})/k_{21}]I_1 k_{21}/(k_{21} - \mu_2)
\end{aligned} \tag{9.10}$$

which are identical to (9.6). The probability density function for the vector of population sizes for any $t > 0$ is therefore

$$p_{x_1,x_2}(t) = \exp\{-\lambda_1(t) - \lambda_2(t)\}[\lambda_1(t)]^{x_1}[\lambda_2(t)]^{x_2}/x_1!x_2!. \tag{9.11}$$

Equation (9.11) may be factored into two marginal Poisson distributions of the form in (3.13), which implies that $X_1(t)$ and $X_2(t)$ are statistically independent random variables [74]. The low-order cumulants for $X_1(t)$ and $X_2(t)$, using Chapter 3 results with the marginal distributions, are

$$\begin{aligned}
\kappa_{10}(t) &= \kappa_{20}(t) = \lambda_1(t) \\
\kappa_{01}(t) &= \kappa_{02}(t) = \lambda_2(t).
\end{aligned} \tag{9.12}$$

The independence of the two variables implies that

$$\kappa_{11}(t) = 0. \tag{9.13}$$

These results could be verified directly by substituting (9.11) into (9.8) and (9.9).

The statistical independence of the population sizes is intriguing. In the context of the roach problem, suppose that the family returns after $t = 10$ weeks. The expected number of roaches from (9.10) and (9.12) are

$$\begin{aligned}
\kappa_{10}(10) &= 100(1 - e^{-1}) = 63.2 \\
\kappa_{01}(10) &= 100[4(1 - e^{-0.5}) - (1 - e^{-1})] = 94.2
\end{aligned}$$

However, suppose that upon their return, the family observes an unusually high number of roaches in the kitchen (area 1), e.g. suppose $X_1(10) = 126$. Though area 2 receives roaches only from area 1, and though the large observed number in area 1 implies a larger total rate of roach migration from 1 to 2, the higher than expected number in 1 does *not* give any information on the population size in 2. In other works, the distribution in area 2 is

Poisson with parameter 94.2 regardless of whether $X_1(10)$ is 32, 63, 126 or any other non-negative integer. This independence property is preserved for any linear n-compartment system with any pattern of interconnections and even for time-varying immigration, provided that the system is initially empty [40] (and, of course, that the roaches do not reproduce during this period).

9.4 Kolmogorov Differential Equations

The use of Kolmogorov differential equations, illustrated for a single population in Chapter 3, extends directly to multiple populations. Jacquez [28, p. 246–248] develops these equations for the general two-compartment model. Many others consider the n-compartment case, including [4, 7, 81, 82], and, in a compartmental context, in [43]. For simplicity, these equations will be illustrated only for the present, restricted model in Figure 9.1.

Generalizing (3.30), let the conditional probabilities of changes in the population sizes from t to $t + \Delta t$ be described using intensity functions, whereby

$$\begin{aligned}&\text{Prob}[X_1(t) \text{ changes by } i \text{ units and } X_2(t) \text{ by } j \text{ units}] \\ &\quad = f_{ij}(X_1, X_2)\Delta t + o(\Delta t).\end{aligned} \tag{9.14}$$

For example, these f_{ij} intensity functions for three possible changes in the present simple model are

$$\begin{aligned} f_{1,0} &= I_1 \\ f_{-1,1} &= k_{21}X_1 \\ f_{0,-1} &= \mu_2 X_2. \end{aligned} \tag{9.15}$$

For present convenience, we denote $X_1(t)$ and $X_2(t)$ as $X(t)$ and $Y(t)$, respectively. In order to obtain the Kolmogorov equations, consider the vector of population sizes at time $t+\Delta t$, say $\mathbf{X}(t+\Delta t) = (x, y)$. Considering first order terms in Δt, there are four mutually exclusive "pathways" from time t to this event at $t + \Delta t$ for the model of interest. These pathways are: i) $\mathbf{X}(t) = (x-1, y)$ with a single immigrant to 1 in Δt, ii) $\mathbf{X}(t) = (x+1, y-1)$ with a single migrant from 1 to 2 in Δt, iii) $\mathbf{X}(t) = (x, y+1)$ with a single death in 2 in Δt, and iv) $\mathbf{X}(t) = (x, y)$ with no change in interval Δt. Therefore, the probability of this event may be written as the sum, using also intensity functions, as

$$\begin{aligned} &p_{x,y}(t + \Delta t) \\ &\quad = p_{x-1,y}(t) f_{1,0}(x-1, y)\Delta t + p_{x+1,y-1}(t) f_{-1,1}(x+1, y-1)\Delta t \\ &\quad + p_{x,y+1}(t) f_{0,-1}(x, y+1)\Delta t \\ &\quad + p_{x,y}(t)\left\{1 - \left[f_{1,0}(x,y) + f_{-1,1}(x,y) + f_{0,-1}(x,y)\right]\Delta t\right\} + o(\Delta t). \end{aligned} \tag{9.16}$$

Substituting (9.15) yields the equation

$$p_{x,y}(t+\Delta t) = p_{x,y}(t)[1-(I_1+k_{21}x+\mu_2 y)\Delta t] + p_{x-1,y}(t)[I_1\Delta t] \qquad (9.17)$$
$$+ p_{x+1,y-1}(t)[k_{21}(x+1)\Delta t] + p_{x,y+1}(t)[\mu_2(y+1)\Delta t] + o(\Delta t),$$

from whence it follows that

$$\dot{p}_{x,y}(t) = -(I_1+k_{21}x+\mu_2 y)p_{x,y}(t) + I_1 p_{x-1,y}(t) \qquad (9.18)$$
$$+ k_{21}(x+1)p_{x+1,y-1}(t) + \mu_2(y+1)p_{x,y+1}(t)$$

for $x \geq 1$ and $y \geq 1$, with appropriate boundary conditions for either $x = 0$ or $y = 0$. The initial conditions are

$$p_{00}(0) = 1, \text{ and} \qquad (9.19)$$
$$p_{i,j}(0) = 0, \qquad \text{for } i + j > 0.$$

It can be shown that the solution in (9.11) satisfies (9.18) and (9.19), and hence that $\mathbf{X}(t)$ is a bivariate Poisson. The solution in (9.11) will be proven subsequently using generating functions.

The Kolmogorov equations in (9.18) can also be expressed using a coefficient matrix $\mathbf{R}$, as in the single population case in (3.17). The $\mathbf{R}$ matrix would be block triangular, but is deferred for present simplicity.

9.5 Bivariate Generating Functions

Convenient methodology for obtaining the joint distributions of random vectors is also available using generating functions. Generalizing the previous development in Chapter 3, consider the integer-valued random vector $\mathbf{X}(t) = [X_1(t), X_2(t)]'$. The probability generating function for this random vector is

$$P(s_1, s_2, t) = \sum_{x_1, x_2} s_1^{x_1} s_2^{x_2} p_{x_1,x_2}(t) \qquad (9.20)$$

where $|s_i| < 1$ for $i = 1, 2$. Clearly, the joint probabilities, $p_{x_1,x_2}(t)$, are the coefficients of powers of s_1 and s_2 in the power series expansion of $P(s_1, s_2, t)$, which will be utilized subsequently.

The moment generating function (mgf) is defined as

$$M(\theta_1, \theta_2, t) = \sum e^{\theta_1 x_1 + \theta_2 x_2} p_{x_1,x_2}(t) \qquad (9.21)$$

where the θ_i are dummy variables. Alternatively, with these non-negative "counting" variables it follows that

$$M(\theta_1, \theta_2, t) = P(e^{\theta_1}, e^{\theta_2}, t). \qquad (9.22)$$

The mgf may be expressed as a power series in θ_1 and θ_2 as

$$M(\theta_1, \theta_2, t) = \sum_{i,j \geq 0} \mu_{ij}(t)\theta_1^i \theta_2^j / i!j!. \qquad (9.23)$$

Hence the joint moment functions may be obtained after proper differentiation, specifically

$$\mu_{ij}(t) = M^{(i,j)}(0, 0, t), \tag{9.24}$$

representing the ith partial with respect to θ_1 of the jth partial with wrt θ_2 evaluated at $s_1 = s_2 = 0$.

The cumulant generating function (cgf) is defined as

$$K(\theta_1, \theta_2, t) = \log M(\theta_1, \theta_2, t) \tag{9.25}$$

with power series expansion

$$K(\theta_1, \theta_2, t) = \sum \kappa_{ij}(t)\theta_1^i\theta_2^j/i!j! \tag{9.26}$$

which formally defines the joint $(i, j)^{th}$ cumulants. The cumulant functions may be obtained in analogous fashion to (9.24) after appropriate differention and evaluation.

As an example, consider the generating functions for the bivariate Poisson distribution in (9.11). Its pgf could be obtained using (9.20) as

$$P(s_1, s_2, t) = \exp\{(s_1 - 1)\lambda_1(t) + (s_2 - 1)\lambda_2(t)\}, \tag{9.27}$$

from whence, using (9.22), the mgf is

$$M(\theta_1, \theta_2, t) = \exp\{(e^{\theta_1} - 1)\lambda(t) + (e^{\theta_2} - 1)\lambda_2(t)\}, \tag{9.28}$$

and, from (9.25), the cgf is

$$K(\theta_1, \theta_2, t) = \sum_i (e^{\theta_i - 1})\lambda_i(t). \tag{9.29}$$

Differentiating with respect to θ_1, it is immediate that the marginal cumulants of $X_1(t)$ are

$$\kappa_{i0}(t) = \lambda_1(t), \quad \text{for } i > 0, \tag{9.30}$$

as given in (9.12). Similarly, one has

$$\kappa_{0j}(t) = \lambda_2(t), \quad \text{for } j > 0. \tag{9.31}$$

The joint cumulant functions from (9.29) are

$$\kappa_{ij}(t) = 0, \tag{9.32}$$

for any combination of $i > 0$ and $j > 0$, which extends (9.13).

9.6 Partial Differential Equations for Generating Functions

Kolmogorov equations such as (9.18) could be transformed directly into partial differential equations for generating functions, as illustrated for a univariate model in Chapter 3. This process is illustrated for an n-dimensional model in [12, 43]. The algebraic manipulations, though not difficult, are tedious, therefore only the "random-variable" technique will be illustrated. The operator equations for the bivariate process, from [4, p. 118–119] are:

$$\frac{\partial P(s_1, s_2, t)}{\partial t} = \sum_{i,j} \left(s_1^i s_2^j - 1\right) f_{ij}\left(s_1 \frac{\partial}{\partial s_1}, s_2 \frac{\partial}{\partial s_2}\right) P(s_1, s_2, t) \quad (9.33)$$

$$\frac{\partial M(\theta_1, \theta_2, t)}{\partial t} = \sum_{i,j} \left(e^{i\theta_1 + j\theta_2} - 1\right) f_{ij}\left(\frac{\partial}{\partial \theta_1}, \frac{\partial}{\partial \theta_2}\right) M(\theta_1, \theta_2, t), \quad (9.34)$$

where, in both expressions, it is understood that the case $i = j = 0$ is excluded in the summation.

Substituting the intensity functions for this model in (9.15) into (9.33) and (9.34) yields

$$\frac{\partial P}{\partial t} = (s_1 - 1)I_1 P + (s_2 - s_1)k_{21} \frac{\partial P}{\partial s_1} + (1 - s_2)\mu_2 \frac{\partial P}{\partial s_2}, \text{ and} \quad (9.35)$$

$$\frac{\partial M}{\partial t} = (e^{\theta_1} - 1)I_1 M + (e^{-\theta_1 + \theta_2} - 1)k_{21} \frac{\partial M}{\partial \theta_1} + (e^{-\theta_2} - 1)\mu_2 \frac{\partial M}{\partial \theta_2}, \quad (9.36)$$

with each possible change in (9.15) leading to a term on the righthand side of the pde. The initial conditions with $X_1(0) = X_2(0) = 0$ from (9.20) and (9.21) are

$$P(s_1, s_2, 0) = M(\theta_1, \theta_2, 0) = 1. \quad (9.37)$$

The linear pde's in (9.35) and (9.36) may be solved using procedures outlined in [12], among others. One could verify by substitution that the solutions are given in (9.27) and (9.28), which prove that the random vector $\mathbf{X}(t)$ for this model follows a bivariate Poisson distribution.

However, as before in Chapter 3, the solution for the bivariate distribution in other cases may be intractable. An alternative approach is to substitute the series expansions of the generating functions, i.e. (9.20) and (9.21), into the pde's such as (9.35) and (9.36) to obtain directly differential equations for the probabilities and for the moments. The former are the Kolmogorov equations, as given in (9.18).

As an example, consider how this approach may be used to obtain differential equations for the cumulant functions. Multiplying both sides of (9.36) by M^{-1} and using the definition of the cgf in (9.25), it follows after simple manipulation that the pde for the cgf is

$$\frac{\partial K}{\partial t} = (e^{\theta_1} - 1)I_1 + (e^{-\theta_1 + \theta_2} - 1)k_{21} \frac{\partial K}{\partial \theta_1} + (e^{-\theta_2} - 1)\mu_2 \frac{\partial K}{\partial \theta_2} \quad (9.38)$$

with initial condition $K(\theta_1, \theta_2, 0) = 0$. Upon substituting the series expansion (9.26) into (9.38), and equating coefficients of θ_1 and θ_2, one has the following differential equations for the mean value functions:

$$\dot{\kappa}_{10}(t) = I_1 - k_{21}\kappa_{10}(t) \tag{9.39}$$
$$\dot{\kappa}_{01}(t) = k_{21}\kappa_{10}(t) - \mu_2\kappa_{01}(t)$$

which are stochastic analogs of the deterministic formulation in (9.5). The equations for the second and third order cumulants, obtain by equating coefficients of second and third order terms of $\theta_1^i\theta_2^j$, are

$$\begin{aligned}
\dot{\kappa}_{20}(t) &= I_1 + k_{21}\kappa_{10} - 2\mu_2\kappa_{20}\\
\dot{\kappa}_{11}(t) &= -k_{21}\kappa_{10} + k_{21}\kappa_{20} - (k_{21} + \mu_2)\kappa_{11} \qquad (9.40)\\
\dot{\kappa}_{02}(t) &= k_{21}\kappa_{10} + \mu_2\kappa_{01} + 2(k_{21} - \mu_2)\kappa_{11}\\
\dot{\kappa}_{30}(t) &= I_1 - k_{21}\kappa_{10} + 3k_{21}\kappa_{20} - 3k_{21}\kappa_{30}\\
\dot{\kappa}_{21}(t) &= k_{21}\kappa_{10} - 2k_{21}\kappa_{20} + (k_{21} + \mu_2)\kappa_{11} + k_{21}\kappa_{30} - (2k_{12} + \mu_2)\kappa_{21}\\
\dot{\kappa}_{12}(t) &= -k_{21}\kappa_{10} + k_{21}\kappa_{20} - (2k_{21} - \mu_2)\kappa_{11} + (2k_{21} - \mu_2)\kappa_{21} - k_{21}\kappa_{12}\\
\dot{\kappa}_{03}(t) &= k_{21}\kappa_{10} - \mu_2\kappa_{01} + 3k_{21}\kappa_{11} + 3k_{21}\kappa_{12}
\end{aligned}$$

which generalize (3.38) to two populations.

One can verify that the solutions to these equations are given by (9.30)–(9.32). The important principle in this discussion is that these solutions for the joint cumulants may be obtained directly from (9.38) without the necessity of first solving the partial differential equation in (9.38) for the cgf in (9.29).

9.7 General Approach to Multiple Population Growth Models

The methodology in this chapter will be used to generalize the single population models in Chapters 3–7 to multiple populations connected by migration. As in these previous chapters, the models of interest will first be illustrated by describing practical applications which extend previous single population applications. The deterministic solutions will then be given, followed by the Kolmogorov equations for the stochastic solution. The partial differential equations for the generating functions will be analyzed, with the primary objective of finding exact or accurate approximations for the cumulant functions. Finally, numerical solutions for the specific applications will be explored.

10

Linear Death-Migration Models

10.1 Introduction

Consider now modeling n populations connected by linear migration, with each population having both immigration and death events occurring at linear rates. Specifically, the assumptions for unit changes in this linear system from t to $t + \Delta t$, using the general formulation in (9.3) and (9.4), would be:

$$\begin{aligned}
&1.\ \text{Prob}\{X_i \text{ will increase by 1 due to immigration }\} = I_i \Delta t \\
&2.\ \text{Prob}\{X_i \text{ will decrease by 1 due to death }\} = \mu_i X_i \Delta t \\
&3.\ \text{Prob}\{X_i \text{ will increase by 1 and } X_j \text{ will decrease by 1 due} \\
&\qquad\qquad \text{to migration}\} \\
&\qquad = k_{ij} X_j \Delta t, \quad \text{for } i \neq j.
\end{aligned} \tag{10.1}$$

This model will be denoted as the LIDM model, where M obviously stands for "migration".

10.2 General Formulation of the Stochastic Model

Consider now the standard stochastic compartmental model, which would include immigration but preclude births. For simplicity, again, only the two-compartment case of this LIDM model will be developed, however the general n-compartment case follows directly, and is fully developed in [43].

In the stochastic formulation, X_1 and X_2 may be regarded as integer-valued population sizes of "particles". The population sizes are assumed to be subject only to unit changes; however the unit change assumption is generalized subsequently. For now, the full set of possible conditional probabilities for the six possible changes, namely an arrival, a death, and a transfer, in each of the two compartments, may be expressed using the intensity functions in (9.14) as

$$\begin{aligned} f_{1,0} &= I_1 & f_{0,1} &= I_2 \\ f_{-1,0} &= \mu_1 X_1 & f_{0,-1} &= \mu_2 X_2 \\ f_{-1,1} &= k_{21} X_1 & f_{1,-1} &= k_{12} X_2. \end{aligned} \tag{10.2}$$

The probability of sizes X_1 and X_2 at $t + \Delta t$ may be written, generalizing (9.16), as

$$p_{x_1,x_2}(t + \Delta t) = \sum_{i,j} p_{x_1-i,x_2-j}(t) f_{i,j}(x_1 - i, x_2 - j)\Delta t \tag{10.3}$$

$$+ p_{x_1,x_2}(t) \left[1 - \sum_{i,j} f_{i,j}(x_1, x_2)\Delta t \right],$$

which yields Kolmogorov equations

$$\dot{p}_{x_1,x_2}(t) = -p_{x_1,x_2} \sum_{i,j} f_{i,j}(x_1, x_2) + \sum_{i,j} p_{x_1-i,x_2-j} f_{i,j}(x_1 - i, x_2 - j), \tag{10.4}$$

where, it is understood that the case $i = j = 0$ is excluded in all summations in (10.3) and (10.4). Upon substituting the specific linear intensity functions in (10.2) into (10.4), the Kolmogorov equation for this general LIDM model is:

$$\begin{aligned} \dot{p}_{x_1,x_2}(t) = {} & -p_{x_1,x_2} [I_1 + I_2 + \mu_1 x_1 + \mu_2 x_2 + k_{21} x_1 + k_{12} x_2] \\ & + p_{x_1-1,x_2} I_1 + p_{x_1,x_2-1} I_2 + p_{x_1+1,x_2} \mu_1 (x_1 + 1) \\ & + p_{x_1,x_2+1} \mu_2 (x_2 + 1) + p_{x_1+1,x_2-1} k_{21} (x_1 + 1) \\ & + p_{x_1-1,x_2+1} k_{12} (x_2 + 1). \end{aligned} \tag{10.5}$$

One could find the coefficient matrix, $\mathbf{R}$, for this system, and then solve the system of equations numerically for the transient probability functions. In fact, that procedure will be followed for the subsequent nonlinear models. For the present linear model, however, there are simpler options based on basic probability arguments for many special cases of the model. One such special case is discussed in this chapter.

For this stochastic development, it is necessary to identify both the origin and the present location of individual particles. Therefore, consider splitting the vector of population sizes, in the present case

$$\mathbf{X}(t) = [X_1, (t), X_2(t)]',$$

into three parts depending on particle origin. These three parts are

$$\mathbf{X}(t) = \mathbf{X}^{(0)}(t) + \mathbf{X}^{(1)}(t) + \mathbf{X}^{(2)}(t), \tag{10.6}$$

where vectors $\mathbf{X}^{(1)}(t)$ and $\mathbf{X}^{(2)}(t)$ denote the numbers of the initial particles which started in compartments 1 and 2, respectively, *at* $t = 0$, and are in the various compartments at time t, and vector $\mathbf{X}^{(0)}(t)$ denotes the numbers of particles which immigrated to the system *after* $t = 0$ and are in the various compartments at time t. This generalizes the partitioning for a one-compartment system in (4.12). The generalization to n compartments is immediate.

This chapter considers the special case where there is no immigration, i.e. where $\mathbf{X}^{(0)}(t) = \mathbf{0}$. This case, denoted as the LDM model, has wide application to experiments based on a pulse-labeling protocol, where labeled "particles" are introduced as a bolus to the system at some initial time, denoted $t = 0$. This special case will be solved directly without utilizing generating functions. The model with immigration, i.e. the LIDM model, will be deferred to Chapter 11, where it will be investigated using the regular generating function approach of this monograph.

10.3 Direct Solution for Stochastic Migration-Death Model

10.3.1 Form of the Probability Distribution for Population Sizes

In order to track both the origin and the present location of individual particles, let

$X_{ij}(t)$ = number of particles starting in j at $t = 0$ which are in i at time t, for $j = 1, \ldots, n$ and $i = 0, \ldots, n$.

The following two-way table represents the population sizes in a two-compartment system as a function of particle origin and location at time t.

		Compartment of Origin		
		1	2	total
Compartment at	1	$X_{11}(t)$	$X_{12}(t)$	$X_1(t)$
time t	2	$X_{21}(t)$	$X_{22}(t)$	$X_2(t)$
	0	$X_{01}(t)$	$X_{02}(t)$	$X_0(t)$
	total	$X_1(0)$	$X_2(0)$	

Note that the initial sizes are $X_j(0) = \sum_i X_{ij}(t)$ for $j = 0, 1$; and that the sizes at time t are $X_i(t) = \sum_j X_{ij}(t)$, for $i = 0, 1, 2$.

The vector of population sizes under the model assumptions which preclude immigration reduces from (10.6) to

$$\mathbf{X}(t) = \mathbf{X}^{(1)}(t) + \mathbf{X}^{(2)}(t), \text{ where} \qquad (10.7)$$
$$\mathbf{X}^{(j)}(t) = [X_{1j}(t), X_{2j}(t)]'$$

for $j = 1, 2$.

Consider also a corresponding table of so-called occupancy probabilities. Let

$\pi_{ij}(t)$ = probability that a particle starting in j at time 0 will be in i at time t, for $j = 1, 2$ and $i = 0, 1, 2$.

A two-way table for these probabilities is

		Compartment of Origin	
		1	2
Compartment at	1	$\pi_{11}(t)$	$\pi_{12}(t)$
time t	2	$\pi_{21}(t)$	$\pi_{22}(t)$
	0	$\pi_{01}(t)$	$\pi_{02}(t)$
	total	1.0	1.0

Consider first the vector $\mathbf{X}^{(1)}(t) = [X_{11}(t), X_{21}(t)]$ in the table, resulting from the initial $X_1(0)$ particles in compartment 1. These $X_1(0)$ particles are assumed to move independently, with identical kinetic flow probabilities, among three mutually exclusive locations, namely compartments 0, 1 and 2. Hence, it follows from basic probability theory [74] that the vector $\mathbf{X}^{(1)}(t)$ would have a trinomial distribution with parameters $X_1(0)$, $\pi_{11}(t)$, and $\pi_{21}(t)$. In symbols, one has

$$\mathbf{X}^{(1)}(t) = [X_{11}(t), X_{21}(t)]' \sim T_1[X_1(0), \pi_{11}(t), \pi_{21}(t)], \tag{10.8}$$

where this trinomial has the formal definition

$$\text{Prob}[X_{11}(t) = x_1, X_{21}(t) = x_2] = X_1(0)!\pi_{11}(t)^{x_1}\pi_{21}(t)^{x_2}\pi_{01}(t)^{x_0}/x_1!x_2!x_0! \tag{10.9}$$

with $x_0 = X_1(0) - x_1 - x_2$ and $\pi_{01}(t) = 1 - \pi_{11}(t) - \pi_{21}(t)$.

Similarly, the counts of particles starting in compartment 2, have the trinomial distribution

$$\mathbf{X}^{(2)}(t) = [X_{12}(t), X_{22}(t)]' \sim T_2[X_2(0), \pi_{12}(t), \pi_{22}(t)], \tag{10.10}$$

with definition analogous to (10.9). Therefore the distribution of the vector of population sizes $\mathbf{X}(t)$ in (10.7) would be the so-called convolution of two trinomial distributions, say

$$\mathbf{X}(t) = [X_1(t), X_2(t)]' \sim T_1[X_1(0), \pi_{11}(t), \pi_{21}(t)] + T_2[X_2(0), \pi_{12}(t), \pi_{22}(t)] \tag{10.11}$$

The marginal distribution for a trinomial distribution, or more generally for any multinomial of n states, is a binomial distribution [74]. Hence, it follows from (10.11) that the marginal distributions of both $X_1(t)$ and $X_2(t)$ would be the convolution of two binomials, say

$$X_1(t) \sim B_1[X_1(0), \pi_{11}(t)] + B_2[X_2(0), \pi_{12}(t)]. \tag{10.12}$$

$$X_2(t) \sim B_3[X_1(0), \pi_{21}(t)] + B_4[X_2(0), \pi_{22}(t)]$$

10.3.2 Occupancy Probabilities

The form of the distributions have been identified in (10.11) and (10.12), but the $\pi_{ij}(t)$ occupancy probabilities remain to be derived as functions of the kinetic parameters. Let

$$\mathbf{\Pi}(t) = [\pi_{ij}(t)] = \text{ an } n \times n \text{ matrix of occupancy probabilities.}$$

One could solve for $\mathbf{\Pi}(t)$ directly by relating the corresponding system of Kolmogorov equations to previous results for the deterministic model, as follows.

Consider first a single random particle that starts in compartment 1. The probabilities $\pi_{11}(t)$ and $\pi_{21}(t)$ would give, respectively, the probabilities that the particle is still in 1 and that the particle has moved to 2 from 1. The derivatives for $\pi_{11}(t)$ and $\pi_{21}(t)$ may be obtained directly from the Kolmogorov equation in (10.5). The correspondences between the $\pi_{ij}(t)$ occupancy probabilities and the $p_{ij}(t)$ multivariate probabilities for the single particle starting in 1 are $\pi_{11}(t) = p_{10}(t)$ and $\pi_{21}(t) = p_{01}(t)$. The equations for $\dot{p}_{10}\ (t)$ and $\dot{p}_{01}\ (t)$ from (10.5), with $I_1 = I_2 = 0$ and under the constraint that x_1, x_2, and $x_1 + x_2$ are all either 0 or 1, are:

$$\begin{aligned} \dot{p}_{10}\ (t) &= -(\mu_1 + k_{21})p_{10}(t) + k_{12}p_{01}(t), \text{ and} \\ \dot{p}_{01}\ (t) &= -(\mu_2 + k_{12})p_{01}(t) + k_{21}p_{10}(t), \end{aligned} \tag{10.13}$$

with all other terms vanishing or inadmissible. Hence it follows that

$$\begin{aligned} \dot{\pi}_{11}\ (t) &= -(\mu_1 + k_{21})\pi_{11}(t) + k_{12}\pi_{21}(t) \\ \dot{\pi}_{21}\ (t) &= k_{21}\pi_{11}(t) - (\mu_2 + k_{12})\pi_{21}(t), \end{aligned} \tag{10.14}$$

with initial conditions $\pi_{11}(0) = 1$ and $\pi_{21}(0) = 0$.

Similarly, for a particle starting in 2 one has $\pi_{12}(t) = p_{10}(t)$ and $\pi_{22}(t) = p_{01}(t)$, from whence by (10.13):

$$\begin{aligned} \dot{\pi}_{12}\ (t) &= -(\mu_1 + k_{21})\pi_{12}(t) + k_{12}\pi_{22}(t) \text{ and} \\ \dot{\pi}_{22}\ (t) &= k_{21}\pi_{12}(t) - (\mu_2 + k_{12})\pi_{22}(t), \end{aligned} \tag{10.15}$$

with initial conditions $\pi_{12}(0) = 0$ and $\pi_{22}(0) = 1$.

In matrix notation, (10.14) and (10.15) may be written as

$$\dot{\mathbf{\Pi}}\ (t) = \mathbf{K}\mathbf{\Pi}(t). \tag{10.16}$$

which is analogous to the deterministic model in (8.4) with the coefficient matrix $\mathbf{K}$. In this stochastic formulation, the initial condition for (10.16) is $\mathbf{\Pi}(0) = \mathbf{I}$.

Though the $\mathbf{K}$ matrix is identical in form in (8.4) and (10.16), it has different interpretations, of course. Conceptually, in the deterministic formulation it describes the proportional *flows* of a unit mass in the various compartments. In the stochastic context, it describes conditional flow *probabilities.*

The solution for the occupancy probabilities, similar to (8.5), is simply

$$\mathbf{\Pi}(t) = \exp(\mathbf{K}t). \tag{10.17}$$

The $\pi_{ij}(t)$ solutions in (8.9) for the general two-compartment deterministic model may therefore also be interpreted as the occupancy probabilities of the general two-compartment stochastic model.

10.3.3 Example of Probability Distribution

As a simple illustration, consider the stochastic model with $\mu_1 = \mu_2 = 2$ and $k_{21} = k_{12} = 1$, which is analogous to the previous deterministic illustration in Section 8.2.3. Consider a "particle" starting in compartment 1 (or because of present symmetry the solution is the same if it had started in 2). The particle may move between compartments many times. However, the probability from (10.9) that it would be in the compartment of origin is, from (8.11):

$$\pi_{11}(t) \text{ or } \pi_{22}(t) = (e^{-2t} + e^{-4t})/2, \tag{10.18}$$

whereas the probability that it is in the other compartment is obviously

$$\pi_{21}(t) \text{ or } \pi_{12}(t) = (e^{-2t} - e^{-4t})/2. \tag{10.19}$$

The likelihood that it has left the system, say $\pi_{01}(t) = 1 - \pi_{ii}(t) - \pi_{ji}(t)$, is

$$\pi_{01}(t) = \pi_{02}(t) = 1 - e^{-2t}. \tag{10.20}$$

To continue this example, suppose one introduces 20 labeled generic "particles" into compartment 1 at $t = 0$, e.g. these could be 20 marked roaches in the application in Section 9.2. The distribution of the roaches remaining in the two compartments at any time t may be obtained from (10.11) with $X_1(0) = 20$ and $X_2(0) = 0$. For example, at $t = 1$, say at the end of the first week, the occupancy probabilities from (10.18)–(10.20) are

$$\begin{aligned} \pi_{11}(1) &= (e^{-2} + e^{-4})/2 = .0768 \\ \pi_{21}(1) &= (e^{-2} - e^{-4})/2 = .0585 \\ \pi_{01}(1) &= 1 - e^{-2} = .8647. \end{aligned}$$

The probability that $X_1(1)$ and $X_2(1)$ take on any specific values, x_1 and x_2, may be obtained by substituting the $\pi_{ij}(1)$ into (10.9). For example, the probability that $x_1 = x_2 = 0$, i.e. of a house free of the initial marked roaches, is

$$\text{Prob}[X_1(1) = X_2(1) = 0] = (0.8647)^{20} = .05457 \tag{10.21}$$

The $X_i(t)$ marginal distributions are given by the binomial distributions in (10.12). For example, consider the probabilities of roach-free individual compartments (or rooms). From the binomial distribution, one has

$$\text{Prob}[X_1(1) = 0] = (1 - .0768)^{20} = 0.20215 \text{ and} \quad (10.22)$$

$$\text{Prob}[X_2(1) = 0] = (1 - .0585)^{20} = 0.29944. \quad (10.23)$$

It is easily shown that the joint probability in (10.21) is not the product of the marginal probabilities in (10.22) and (10.23). Therefore, $X_1(t)$ and $X_2(t)$ are *not* independent, as they were in the simple example in Section 9.2 with immigration but with no initial roaches.

Such probability distributions are often of interest in population biology applications, where for example one may want to manage the size of an animal population. In such cases, the population sizes are often relatively small; numbers such as nearly 100,000 muskrats harvested annually in certain provinces in Section 2 are on the large side of typical applications. However, such is not the case in many physiological and pharmacokinetic applications, as noted by Jacquez [28, p.244]. In such applications, the "particles" represent individual molecules, with over 10^{20} molecules per mole. The theory still holds, but the counting distribution might not be of practical interest.

10.3.4 Cumulants for Population Size

The cumulants of $X_1(t)$ are easily obtained using (10.12) with the binomial properties initially introduced in Section 4.5. The first three cumulants are

$$\begin{aligned} \kappa_{10}(t) &= \Sigma X_i(0)\pi_{1i} \\ \kappa_{20}(t) &= \Sigma X_i(0)\pi_{1i}(1 - \pi_{1i}) \\ \kappa_{30}(t) &= \Sigma X_i(0)\pi_{1i}(1 - \pi_{1i})(1 - 2\pi_{1i}), \end{aligned} \quad (10.24)$$

The cumulants $\kappa_{01}(t)$, $\kappa_{02}(t)$ and $\kappa_{03}(t)$ for $X_2(t)$ may be found from (10.24) by replacing π_{1i} by π_{2i}.

The previous subsection indicated the statistical dependency between $X_1(t)$ and $X_2(t)$. The covariance follows from (10.11), using properties of the trinomial, [74], as:

$$\kappa_{11}(t) = -[X_1(0)\pi_{11}(t)\pi_{21}(t) + X_2(0)\pi_{12}(t)\pi_{22}(t)], \quad (10.25)$$

which obviously is always negative.

As an illustration of these cumulants, consider again the previous example of a stochastic model with $X_1(0) = 20$. The mean value functions, using (10.18)–(10.20) and (10.24) are

$$\begin{aligned} \kappa_{10}(t) &= 10(e^{-2t} + e^{-4t}) \text{ and} \\ \kappa_{01}(t) &= 10(e^{-2t} - e^{-4t}). \end{aligned} \quad (10.26)$$

The variance functions are

$$\begin{aligned} \kappa_{20}(t) &= 5(e^{-2t} - 2e^{-6t} - e^{-8t}) \text{ and} \\ \kappa_{02}(t) &= 5(e^{-2t} - 2e^{-4t} + 2e^{-6t} - e^{-8t}), \end{aligned} \quad (10.27)$$

and the covariance function is

$$\kappa_{11}(t) = -5(e^{-4t} - e^{-8t}). \tag{10.28}$$

As noted previously, the usefulness of the stochastic model has been questioned for applications involving the astronomical population sizes envisioned in many physiological and pharmacokinetic systems [28]. It is clear from (10.24), as noted for equation (4.26) and observed in [28, p. 244], that the coefficient of variation for any population size approaches 0 as the $X_i(0)$ increase. The logical argument follows that perhaps the deterministic model is hence sufficient for linear systems with such large population sizes. A discussion of when the stochastic model may be useful even in such cases of enormous population sizes is given in [28, Section 12.2]. However, in our view the most compelling overall practical reason for using the stochastic model, for large or for small populations, is outlined in the next section.

10.4 Mean Residence Times

10.4.1 Calculation of Mean Residence Times

One great advantage of the stochastic model, in addition to its theoretical appeal of a broader conceptual foundation, is that it leads to additional methodology of interest for describing system dynamics. In particular, many new random variables may be defined for a stochastic multicompartment model and then utilized to describe the dynamics of particle transfer among the various compartments. Some of the more useful variables are as follows. Let

1. R_i denote the retention time of a particle during a single visit in i, for $i = 1, 2, \ldots n$;
2. N_{ij} be the number of visits that a particle starting in j will make to i prior to its exit from the system, for $i, j = 1, \ldots, n$; and
3. S_{ij} be the total residence time that a particle originating in j will accumulate in i during all of its N_{ij} visits, i.e., $S_{ij} = \sum_1^{N_{ij}} R_i$.

The concept of mean residence times, or MRT, is a relatively new characterization of great practical utility in a linear compartmental system [28]. Let

4. $E(S_{ij})$ denote the mean of S_{ij}, and
5. $\mathbf{E}(S)$ denote the $n \times n$ matrix of MRT, with elements $E(S_{ij})$.

This section discusses the use of this one measure. The means and variances of other residence time measures, as well as of the related N_{ij} variables which describe particle cycling, are given in [41, 50, 60].

The key result for the MRT of a compartmental system is:

Theorem:

The mean residence times (MRT) for the system of compartments are

$$\mathbf{E}(S) = -\mathbf{K}^{-1}. \tag{10.29}$$

The proof of this key result is not difficult [60]. Consider a so-called indicator variable, say $I_{ij}(t)$, that a particle starting in j is in i at time t. The variable is defined such that $I_{ij}(t) = 1$ when the particle is in i, with $I_{ij}(t) = 0$ otherwise. The residence time in i is then the integral of $I_{ij}(t)$ over t, and the mean residence time, MRT, is the expected value of this integral. Equivalently, the MRT may be obtained as the integral of the expected value of $I_{ij}(t)$. Because the expected value of $I_{ij}(t)$ is $\pi_{ij}(t)$, the MRT is the integral of $\pi_{ij}(t)$. Combining all MRT using matrix notation, one has

$$\mathbf{E}(S) = \int_0^\infty \mathbf{\Pi}(t) = \int_0^\infty \exp(\mathbf{K}t) = -\mathbf{K}^{-1} \tag{10.30}$$

This result on MRT is widely used, in large part because the mean times spent at particular sites are often the natural response variables of interest, instead of the k_{ij} turnover rates of the system. This point will be illustrated subsequently. The fact that the calculations are also simple, involving only matrix inversion in (10.29), adds to the popular appeal of the MRT.

10.4.2 A Simple Example

The use of the MRT may be illustrated with the roach movement model in Section 10.3.3. Assuming $\mu_1 = \mu_2 = 2$ and $k_{21} = k_{12} = 1$, the coefficient matrix $\mathbf{K}$ is

$$\mathbf{K} = \begin{bmatrix} -3 & 1 \\ 1 & -3 \end{bmatrix}.$$

The MRT are given by

$$\mathbf{E}(S) = -\mathbf{K}^{-1} = \begin{bmatrix} .375 & .125 \\ .125 & .375 \end{bmatrix},$$

as obtained also in a different context for equilibrium storages in Section 10.2.2. For example, note that a roach starting in comp 1 is expected to spend 0.375 time units in comp 1, 0.125 in comp 2, and hence 0.500 in the total system prior to departure. These seem consistent with the result in (10.20); in particular at time 1 one has $\pi_{01}(1) = 0.8647$, which indicates a high probability that a random roach has departed the system by time 1.0.

10.4.3 Application to Calcium Kinetics Problem

The MRT concept has been applied in many contexts, including the previous example of mercury accumulation in fish [57] and a number of examples of animal migration, such as [24]. We consider now applying the concept to the previous example of calcium kinetics given in Section 8.3.2. The estimated coefficient matrix **K**, obtained by substituting the estimated coefficients into (8.23), is

$$\mathbf{K} = \begin{bmatrix} -3.0652 & 2.999 & 0.162 \\ 2.611 & -2.999 & 0 \\ 0.389 & 0 & -0.162 \end{bmatrix}.$$

The MRT matrix follows from (10.30) as

$$\mathbf{E}(S) = \begin{bmatrix} 15.344 & 15.344 & 15.344 \\ 13.362 & 13.359 & 13.692 \\ 36.860 & 43.038 & 36.858 \end{bmatrix}. \qquad (10.31)$$

In this example, the first column of $\mathbf{E}(S)$ gives the MRT for a calcium particle starting in compartment 1, the plasma. Its MRT in plasma, soft tissue, and bone are 15.344, 13.362 and 36.860 hrs, respectively, for an estimated mean of 50.22 hrs in the body tissue prior to the particle elimination from the body. This application illustrates that the primary endpoints of interest may not be the estimated rate coefficients, k_{ij}, nor the estimated occupancy probability, $\pi_{11}(t)$, obtainable from (8.24). Instead the desired endpoint may be the expected lifetime of a calcium particle in the body tissue. This measure has been shown to be helpful in classifying the medical condition of patients [20, 21].

In this and similar practical applications, the k_{ij} coefficients are estimates, with estimated standard errors. It follows that the MRT from (10.30) are also estimates. The estimated standard errors for the MRT may be obtained using statistical theory associated with transformed variables [2]. Often the standard errors of the MRT are smaller relatively than the standard errors of the estimated k_{ij} rate coefficients; as a result of the high statistical multicollinearity of the latter. Consequently, in designed experiments the estimated MRT's may be statistically more powerful response variables than the corresponding k_{ij}'s.

As an example, data from a completely randomized design used to test the efficacy of two drugs for reducing cholesterol in mice were analyzed in [60]. Each drug was given to 15 experimental units (mice). Clearance data were obtained, and the rate coefficients for the three compartment model were estimated for each mouse. Treatment differences were observed for all of the five unique mean residence time end points, but for only two of the five individual rate coefficient end points.

10.5 Appendix

10.5.1 An Alternative Model Formulation Based on Retention Time Concepts

10.5.1.1 General Definitions and Concepts

Heretofore, the stochastic models have been conceptualized based on conditional flow probabilities, such as assumptions of unit changes in (3.2), (9.3) and (10.1). Such flow probabilities are natural analogs to the assumed flow proportions of the corresponding deterministic models. However, stochastic models may also be conceptualized based on the time intervals between events, which are called the "retention" (or alternatively transit or sojourn) times of particles.

This section discusses first the equivalent formulation of the linear death-migration models using exponential retention times. The subsequent generalization to nonexponential retention times provides a rich family of extensions for the stochastic models. Such extensions are very useful in practice and yet are *not* readily conceptualized for the corresponding deterministic models.

Consider the following notation. Let

1. R be the retention time of a random particle in a particular compartment.

2. $F(a)$ =Prob $[R \leq a]$ be the distribution function of R, i.e. the probability that the particle will leave the compartment prior to elapsed time, or "age", a,

3. $f(a) = dF(a)/da$ be the probability density function of R,

4. $S(a) = \text{Prob}[R > a] = 1 - F(a)$ be the "survivorship function" of R, i.e. the probability that the particle in the compartment "survives" to age a,

5. $\lambda(a) = f(a)/S(a)$ be the "hazard" rate, defined by the conditional probability:

$$\text{Prob [particle leaves by age } a + \Delta a \text{ given it survives to age } a] = \lambda(a)\Delta a \tag{10.32}$$

As an example, let R have an exponential distribution with parameter k, denoted $R \sim \exp(k)$. By definition, one has

$$f(a) = k\exp(-ka),$$

from whence it follows that

$$\begin{aligned} F(a) &= 1 - \exp(-ka), \\ S(a) &= \exp(-ka), \text{ and} \\ \lambda(a) &= k. \end{aligned} \tag{10.33}$$

This property of a constant (i.e. age–invariant) hazard rate in (10.33) characterizes the exponential distribution [78]. Therefore, particles with exponential retention times are said to "lack a memory" regarding their accumulated elapsed time, i.e. their age, in the compartment. In practical applications, the transfer mechanism for particle elimination from such an 'exponential' compartment must not discriminate on the basis of the age of the particle.

Suppose one assumes that there are $X(t)$ *independent* particles in the compartment, possibly of varying age, but each with retention time variable $R \sim \exp(k)$. According to (10.32) and (10.33), each particle would have independent probability $k\Delta t$ of leaving in the next time increment, Δt. Therefore, from the binomial distribution, the probability of a single departure is:

$$\begin{aligned} &\text{Prob}\{\text{ exactly one of the } X(t) \text{ particles leaves in } \Delta t\} \\ &= \binom{X(t)}{1}(k\Delta t)^1(1 - k\Delta t)^{X(t)-1} \\ &= kX(t)\Delta t + o(\Delta t) \qquad (10.34) \end{aligned}$$

where $o(\Delta t)$ denotes terms of degree 2 or more with respect to Δt. The $o(\Delta t)$ terms vanish in formulating the Kolmogorov differential equations, as in Section 10.2. Therefore, the assumption of independent particles, each with an identical exponential retention time, which yields (10.34), is equivalent to the linear kinetic formulation with constant rates, as for example, given in (10.1).

10.5.1.2 Semi-Markov Models and Erlang Retention Time Distributions

Stochastic systems in which all the compartments have exponential retention times, with consequential constant hazard rates as in (10.33), are said to follow a Markov process [43, 49, 50, 60, 100]. If one or more of compartments had a nonexponential retention time, (with the condition that when transfers do occur, they occur with fixed probabilities among the compartments), the stochastic system would be described by a semi-Markov process. Such models are generally just classified as "non-Markovian" models [100].

In many important applications, the passage probability in (10.32) has an age-varying hazard rate, which necessitates a semi-Markov model. In ruminant kinetics, for example, the passage probability of a particle increases with its time spent in the compartment [15, 16]. The Erlang retention time distributions are widely used in such ruminant nutrition modeling, where it is called age-dependent turnover [39].

Suppose R follows an Erlang distribution with "shape parameter" integer n and "scale" parameter k, denoted $R \sim E(n,k)$. By definition, one has

$$f(a) = k^n a^{n-1} \exp\{-ka\}/(n-1)! \quad n = 1, 2, \ldots,$$

which obviously is an exponential distribution for $n = 1$. One could show

$$F(a) = 1 - \sum_{j=0}^{n-1} (ka)^j \exp\{-ka\}/j!,$$

from whence the hazard rate follows as

$$\lambda(a) = \frac{k(kt)^{n-1}/(n-1)!}{\sum_{j=0}^{n-1}(kt)^j/j!}.$$

Note that, for $n > 1$, the hazard rate $\lambda(a)$ is an increasing function, with asymptotic value k. Figure 10.1 illustrates some hazard rates from Erlang distributions for some small n with $k = 1$. Erlang distributions with $n = 2$ and $n = 3$ have been frequently used to describe ruminant passage/digestion data [15, 16].

Figure 10.1. Hazard rates for four Erlang distributions, $n = 1$ to 4, with common $k = 1$.

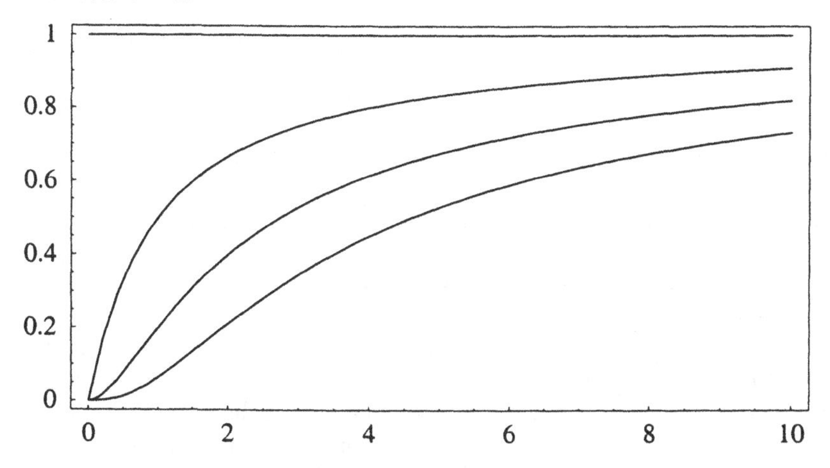

The Erlang distributions are widely used because, in addition to yielding the desired shape of the hazard rates, they are also convenient to generate using the LDM model. One can show [34, 39, 49] that an Erlang (n, k) variable may be obtained as the sum of n independent exponential variates, each with parameter k, *i.e.*,

$$R_i \sim \exp(k), \text{ for } i = 1, \ldots, n,$$

$$\sum_{i=1}^{n} R_i \sim \text{Erlang}(n, k).$$

The practical utility of this result is that one may generate the retention time of an "age-dependent" compartment with, say, an Erlang $(2, k)$ retention time distribution as the retention time for a sequence of two exponential subcompartments, each with rate k. The exponential subcompartments are sometimes called "pseudo-compartments", which have no physical (i.e. mechanistic) interpretation but are merely a mathematical artifice to generate the desired retention times. The use of Erlang retention times is not illustrated here, however besides the application to ruminant nutrition, their use has been illustrated in other animal migration problems in [61], and they will also be illustrated subsequently in Chapter 12.

10.5.1.3 Phase-Type Retention Time Distributions

A general approach for utilizing the semi-Markov structure to incorporate non-exponential retention times into our stochastic models is to utilize phase-type (or PH) distributions, as recently developed by Neuts [71]. A PH distribution is a distribution that may be obtained as the time to departure (or absorption) in a finite LDM model. Mathematically, the survival function, $S(t)$, of a PH distribution could be constructed as

$$S(t) = (B_1, \ldots, B_n) \exp(\mathbf{K}t) \begin{pmatrix} 1 \\ 1 \\ \vdots \\ 1 \end{pmatrix}$$

where the B_i weights and the k_{ij} elements of the transition matrix $\mathbf{K}$ are parameters of the model. In particular, the Erlang $(2, k)$ distribution is a simple PH distribution, as it may be represented as the time to departure from two sequential, identical compartments.

The PH distributions are a very useful tool in the present context due to the following result from [71]:

Claim: Any nondegenerate distribution of a positive variable may be represented as a PH distribution.

The utility of this result is that, in principle, any realistic retention time distribution could be represented using a compartmental subsystem. In theory, this enables one to solve any semi-Markov model with any nondegenerate retention time distributions as an expanded Markovian model with exponential retention times. Of course, the theorem only guarantees the existence of such a compartmental representation. The actual matching of an arbitrary retention time distribution to an exact or suitable approximate PH distribution is a formitable problem, though a procedure based on cumulant matching is suggested in [31]. Fortunately, knowing that the PH representations exist, it is often sufficient in practice to obtain suitably close approximations using data fitting procedures, as illustrated in the subsequent example.

10.5.1.4 Application to Human Calcium Kinetics

We consider again the problem of modeling calcium kinetics, using the example introduced in Section 8.3.2. A group of scientists, including Wise [102], has questioned for many years the use of linear compartmental modeling for the clearance of some elements, particularly calcium, from the human body. Their chief criticism is the implied use of exponential retention times, which requires that all particles have the same passage probabilities, regardless of their ages in the compartments of interest. This in turn would imply, physiologically, a so-called "homogeneous", "well-mixed" compartment. Such assumption is not physiologically realistic for many substances, including in particular the calcium in the bone compartment.

We have previously proposed using Erlang retention times to provide a more realistic model for calcium clearance [49]. More recently, a general modeling approach has been formulated based on PH distributions in [51]. In brief, the approach is first to conceptualize the model in terms of physiological compartments, possibly with nonexponential retention times. An expanded linear system utilizing approximating PH distributions would then be formulated and solved.

In applying this approach to the calcium clearance problem, compartments 1 and 2, representing the plasma and soft tissue, may reasonably be regarded as exponential compartments. However the third compartment, representing the hard tissue including bone, could not from a physiological view point be regarded as well-mixed and homogeneous.

Consider instead, representing compartment 3 using a PH distribution represented in Figure 10.2, which is a sequence of two different Erlang (2) distributions. This system of three phenomenological compartments could then be represented as a six compartment (Markovian) LDM model. The number of parameters in the expanded model remains at six, the same number as in the original Markovian model in Section 8.3.2 with exponential compartments.

Details of the parameter estimation, including the standard errors of the parameters, are given in [51]. The estimated coefficient matrix, $\mathbf{K}^*$, of the expanded system is

$$\mathbf{K}^* = \begin{bmatrix} -3.6357 & 3.131 & 0 & 0.414 & 0 & 0 \\ 3.131 & -3.735 & 0 & 0 & 0 & 0 \\ 0.444 & 0 & -0.444 & 0 & 0 & 0 \\ 0 & 0 & 0.444 & -0.444 & 0 & 0.030 \\ 0 & 0 & 0 & 0.030 & -0.030 & 0 \\ 0 & 0 & 0 & 0 & 0.030 & -0.030 \end{bmatrix}$$

This matrix has complex eigenvalues, as almost always the case in systems involving substantial cycling. Therefore, the fitted curve for the data involves damped oscillations; specifically the equation is given by *Kinetica* is

$$\begin{aligned} E[X_{11}(t)] &= 356.16\exp(-7.105t) + 68.50\exp(-0.0078t) \\ &\quad + \exp(a_1 t)[44.98\sin(b_1 t) + 187.85\cos(b_1 t)] \\ &\quad + \exp(a_2 t)[32.85\sin(b_2 t) + 110.26\cos(b_2 t)] \end{aligned}$$

where $a_1 = -0.5659$, $a_2 = -0.0380$, $b_1 = 0.2852$ and $b_2 = 0.0162$.

Figure 10.2. Schematic of the age-dependent three-compartment model of calcium clearance. The rate k_x is defined as $k_x = k_{13} + k_{31}$. The four pseudo-compartments constituting compartment 3 provide a PH retention time distribution.

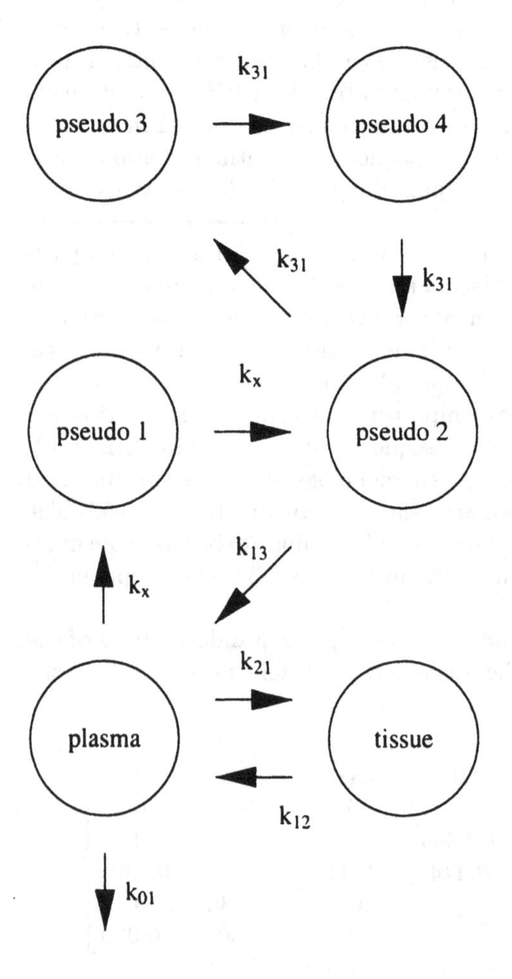

The fitted curve is illustrated with the data points in Figure 10.3. Note that the residuals are tiny, and apparently randomly scattered, at least until $t = 300$, after which time they are virtually negligible. Consequently, the mean squared error for this model is only MSE=4.94, as compared to MSE=27.57 for the homogeneous model in Section 8.3.2. The matrix of estimated MRT follows from (10.29) by simple matrix inversion of $\mathbf{K}^*$ as

$$\mathbf{E}^*[S] = \begin{bmatrix} 16.469 & 16.469 & 16.469 & 16.469 & 16.469 & 16.469 \\ 13.805 & 14.073 & 13.805 & 13.805 & 13.805 & 13.805 \\ 16.469 & 16.469 & 18.718 & 16.469 & 16.469 & 16.469 \\ 17.668 & 17.668 & 20.081 & 20.081 & 20.081 & 20.081 \\ 17.668 & 17.668 & 20.081 & 20.081 & 53.230 & 20.081 \\ 17.668 & 17.668 & 20.081 & 20.081 & 53.230 & 53.230 \end{bmatrix}.$$

The MRT for the plasma and soft tissue compartments of the new model are 16.5 and 13.8, respectively, which are close to the estimates of 15.3 and 13.4 from the homogeneous model. However the MRT for bone is the sum of the MRT in the four pseudo-states in the rest of the first column, i.e. MRT=16.5+3(17.7) = 69.5 This estimate is nearly a 100% increase from the estimate MRT=36.9 from the alternative model.

It seems remarkable that the estimated MRT in bone would differ so drastically in light of the fact that both curves fit the data quite well. Conceptually, the substantial increase in the MRT for the bone is attributable to the improved fit in the far right tail of the clearance data. An assumed exponential bone compartment, even with a small parameter k, is not sufficiently flexible to generate the desired shape characteristic of a long tail. However the fitted PH retention time distribution has the desired long tail property. This feature is generated conceptually by its second Erlang distribution, which adds a long retention time to the small proportion of particles which are expected to experience that second Erlang cycle [51].

The numerical results in this section are based on the PH distribution illustrated conceptually in Figure 10.2. The actual shape and analytical expression for this PH distribution are given in [51] using a linear compartment analysis for this subsystem only. Although this particular assumed PH model obviously provides an excellent fitting of the data, there is no assurance that it is either unique nor the best. Indeed recent results [106] indicate that a similar PH model with *five* pseudo-compartments improves the goodness-of-fit slightly, but the use of *six* pseudo-compartments apparently does not. The results are the same qualitatively when weighted nonlinear least squares is used with the observations as weights.

Figure 10.3. Observed data and fitted curve from the age-dependent three-compartment stochastic model of calcium clearance.

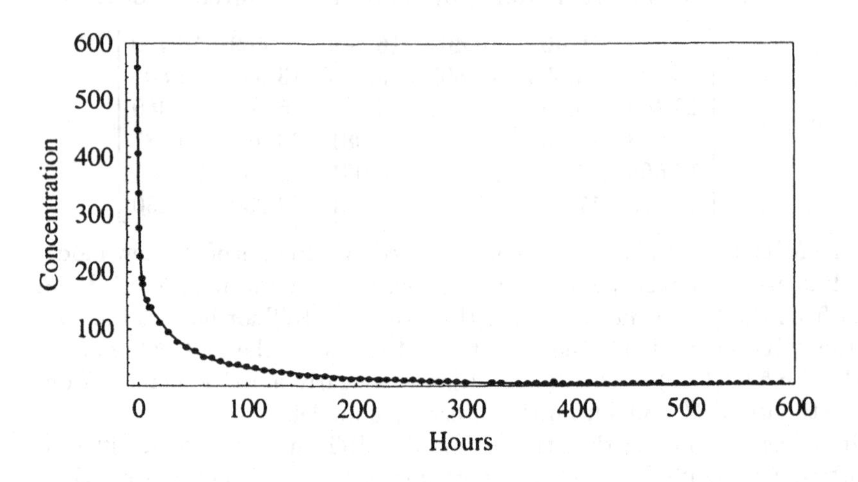

10.5.1.5 Contributions of the Semi-Markov Model Concept

In a broader way, the modeling approach based on general nonexponential (PH) retention times is based on a statistical foundation which, in turn, is implemented using the stochastic LDM model. It is difficult to envision a simple analogous formulation using deterministic models. For example, adding a fourth exponential compartment is impractical, due to statistical multicollinearity problems [8]. The only alternative would seem to be proposing some detailed, deterministic mechanistic model, which to the authors' knowledge has not been proposed for analyzing such calcium clearance data.

In summary, we suggest that the statistical concept of nonexponential retention times is an attractive advantage of stochastic compartmental modelling. As this example illustrates, the use of nonexponential retention times leads to a new class of models which are often better descriptions (or caricatures) of the underlying biological system. The new models often provide an improved fit of the observational data, typically without adding parameters explicitly, and more importantly they may yield more realistic estimates of such biologically useful parameters as the MRT.

11

Linear Immigration-Death-Migration Models

11.1 Introduction

Consider now the stochastic LIDM model, which generalizes the basic single-population LID model introduced in Chapter 4. A simple case of this LIDM model is considered in Chapter 9 to illustrate the basic methodology for multiple populations. This chapter develops a more general case, specifically the case with multiple inputs and two-way migration between compartments. The corresponding deterministic model is given in (8.4) of Chapter 8, with its solution in (8.5) and (8.6), and with the illustration of mercury uptake by fish in Section 8.3.1.

11.2 Generating Functions for the Stochastic Model

The special case of $n = 2$ is sufficient again to illustrate the basic concepts, hence we consider it in some detail and merely indicate the solution for the case of general n. The intensity functions for this model, as previously given in (10.2) are

$$
\begin{aligned}
f_{1,0} &= I_1 & f_{0,1} &= I_2 \\
f_{-1,0} &= \mu_1 X_1 & f_{0,-1} &= \mu_2 X_2 \\
f_{-1,1} &= k_{21} X_1 & f_{1,-1} &= k_{12} X_2.
\end{aligned}
\tag{11.1}
$$

The Kolmogorov equation is given in (10.5), which was solved directly for the case without immigration in Chapter 10.

Consider now the partial differential equations for the generating functions. By substituting (11.1) into the operator equation in (9.33), one has

$$
\begin{aligned}
\frac{\partial P(s_1, s_2, t)}{\partial t} &= P\Sigma(s_i - 1)I_i + [(1 - s_1)\mu_1 + (s_2 - s_1)k_{21}]\frac{\partial P}{\partial s_1} \\
&\quad + [(1 - s_2)\mu_2 + (s_1 - s_2)k_{12}]\frac{\partial P}{\partial s_2},
\end{aligned}
\tag{11.2}
$$

which generalizes (9.35). Similarly, substitution into (9.34) with the transformation to the cumulant generating function in (9.25) yields

$$\frac{\partial K(\theta_1,\theta_2,t)}{\partial t} = \Sigma(e^{\theta_i}-1)I_i + \left[(e^{-\theta_1}-1)\mu_1 + (e^{-\theta_1+\theta_2}-1)k_{21}\right]\frac{\partial K}{\partial \theta_1} + \left[(e^{-\theta_2}-1)\mu_2 + (e^{\theta_1-\theta_2}-1)k_{12}\right]\frac{\partial K}{\partial \theta_2}. \tag{11.3}$$

The solutions to these, under the general condition that $X_1(0) = X_1$ and $X_2(0) = X_2$, are

$$P(s_1,s_2,t) = \exp\left\{\sum_{i=1}^{2}(s_i-1)\alpha_i(t)\right\}\prod_{i=1}^{2}\left[1+\sum_{j=1}^{2}(s_j-1)\pi_{ji}(t)\right]^{X_i} \tag{11.4}$$

and

$$K(\theta_1,\theta_2,t) = \sum_{i=1}^{2}(e^{\theta_i}-1)\alpha_i(t) + \sum_{i=1}^{2}X_i ln\left[1+\sum_{j=1}^{2}(e^{\theta_i}-1)\pi_{ji}(t)\right] \tag{11.5}$$

where the $\pi_{ji}(t)$ parameter functions are the occupancy probabilities described in Section 10.3.2 and given explicitly in (8.9) and (8.10), and the $\alpha_i(t)$ parameter functions are identical to the corresponding functions for the deterministic model in (8.5). These will be illustrated subsequently.

11.3 Probability Distribution

11.3.1 Solution for Probability Distribution

The significance of solutions (11.4) and (11.5) is that they are the product and (equivalently) the sum, respectively, of three terms. This fact implies that the vector of population sizes may be partitioned into three independent random vectors, given in (10.6) as

$$\mathbf{X}(t) = \mathbf{X}^{(0)}(t) + \mathbf{X}^{(1)}(t) + \mathbf{X}^{(2)}(t). \tag{11.6}$$

The vector $\mathbf{X}^{(0)}(t)$, which denotes the numbers of particles which have arrived after $t = 0$ but which are still in the various compartments at time t has a multivariate Poisson distribution, with density function given in (9.11). The vectors $\mathbf{X}^{(1)}(t)$ and $\mathbf{X}^{(2)}(t)$, which denote the particles which originated at $t = 0$ in compartments 1 and 2 respectively, have the trinomial distributions derived in Section 10.3. Symbolically,

$$\begin{aligned} \mathbf{X}^{(0)}(t) &\sim MV\text{Poisson}[\alpha_1(t),\alpha_2(t)] \\ \mathbf{X}^{(1)}(t) &\sim T\left[X_1(0),\pi_{11}(t),\pi_{21}(t)\right] \\ \mathbf{X}^{(2)}(t) &\sim T\left[X_2(0),\pi_{12}(t),\pi_{22}(t)\right] \end{aligned} \tag{11.7}$$

The equilibrium distribution is often of particular interest. Because the $\pi_{ji}(t)$ are sums of negative exponential terms, they vanish in the limit as $t \to \infty$. One is left with the multivariate Poisson distribution with the equilibrium parameters

$$\alpha_1^* = \left[I_1(\mu_2 + k_{12}) + I_2 k_{12}\right]/d \text{ and} \qquad (11.8)$$

$$\alpha_2^* = \left[I_1 k_{21} + I_2(\mu_1 + k_{21})\right]/d$$

where

$$d = (\mu_1 + k_{21})(\mu_2 + k_{12}) - k_{12}k_{21}.$$

The parameters α_1^* and α_2^* are equal to the deterministic equilibrium values X_1^* and X_2^* given in (8.13). Clearly, however, the solution to the stochastic model in equilibrium is a multivariate distribution as compared to the single point solution of the previous deterministic model.

11.3.2 Example of Probability Distribution

In practice, an experimenter would usually follow one of two protocols. In the first, an initial population would be marked, at $t = 0$, and data describing the "wash-out" of these initial units would be obtained. The kinetic parameters of various models could be estimated as the models are fitted to data, as illustrated for the calcium kinetics application in Sections 8.3.2, 10.4.3 and A10.1. The distributions generated by this protocol were illustrated with, among others, the roach migration example in Section 10.3.3.

Consider now the other protocol, where the system initially has no marked individuals, and instead marked immigrants arrive after $t = 0$. Typically data on the build-up of such immigrants would be recorded. Once again, relevant kinetic parameters could be estimated as models are fitted to such data, as illustrated for the mercury bioaccumulation application in Sections 4.7.1 and 8.3.1. For simplicity, the example below will illustrate the distributions generated by this protocol by extending the previous roach example. In particular, suppose that no roaches (at least no marked roaches) are present at $t = 0$ and instead they arrive after $t = 0$ as immigrants. The case of possible (marked) initial *and* immigrant particles, which is usually avoided in practice, is not of present interest but could easily be obtained by combining results for the two individual models.

Consider the two-compartment system with rates $\mu_1 = \mu_2 = 2$ and $k_{21} = k_{12} = 1$, as previously described in Sections 8.2.3 and 10.3.3. Assuming that there are no initial units, i.e., $X_1(0) = X_2(0) = 0$, let instead roaches enter the system, into compartment 1 only, at the rate of $I_1 = 10/\text{wk}$, similar to the illustration in Section 9.2.

The solution for these particular parameter values follows easily using previous results. The vector of inputs is

$$\mathbf{I} = [10, 0]',$$

from whence upon substituting the occupancy probabilities in (8.11) into the matrix exponential (8.8), one may obtain the vector product

$$\exp[\mathbf{K}(t-s)]\mathbf{I} = \begin{bmatrix} 5(e^{-2(t-s)} + e^{-4(t-s)}) \\ 5(e^{-2(t-s)} - e^{-4(t-s)}) \end{bmatrix}. \tag{11.9}$$

Following (8.5), the parameter functions $\alpha_1(t)$ and $\alpha_2(t)$ are the integrals of (11.9) with respect to s, i.e.

$$\begin{aligned} \alpha_1(t) &= 1.25(3 - 2e^{-2t} - e^{-4t}) \\ \alpha_2(t) &= 1.25(1 - 2e^{-2t} + e^{-4t}) \end{aligned} \tag{11.10}$$

Clearly, the equilibrium parameter values are

$$\alpha_1^* = 3.75 \quad \text{and} \quad \alpha_2^* = 1.25$$

as one could also obtain by direct substitution into (11.8). The multivariate Poisson for these values, which is the equilibrium distribution for this roach model, is illustrated in Figure 11.1.

11.4 Cumulant Functions

The cumulant functions for the present LIDM model could be obtained in several ways, as in Chapter 4. Using the known properties of the trinomial and bivariate Poisson distributions in (11.7), it follows directly that

$$\begin{aligned} \kappa_{10}(t) &= X_1(0)\pi_{11} + X_2(0)\pi_{12} + \alpha_1 \\ \kappa_{01}(t) &= X_1(0)\pi_{21} + X_2(0)\pi_{22} + \alpha_2 \\ \kappa_{20}(t) &= X_1(0)\pi_{11}(1-\pi_{11}) + X_2(0)\pi_{12}(1-\pi_{12}) + \alpha_1 \\ \kappa_{11}(t) &= -X_1(0)\pi_{11}\pi_{21} - X_2(0)\pi_{21}\pi_{22} \\ \kappa_{02}(t) &= X_1(0)\pi_{21}(1-\pi_{21}) + X_2(0)\pi_{22}(1-\pi_{22}) + \alpha_2 \end{aligned} \tag{11.11}$$

with π_{ij} and α_i as previously given (with suppressed time variable, t). Extensions to higher order cumulant follow using properties of the trinomial distribution, and of the multivariate Poisson, where as given in (9.30)–(9.32), the cumulants are equal for each marginal distribution and the joint cumulants are 0. A closely related derivation of these would be the direct expansion and subsequent differentiation of the cumulant generating function in (11.5).

Also, as before, it is possible to expand the pde's as a power series to obtain differential equations for the probabilities and for the cumulants.

Figure 11.1. Multivariate Poisson equilibrium distribution of population sizes for LIDM model with parameters $I_1 = 10$, $\mu_1 = \mu_2 = 2$, and $k_{21} = k_{12} = 1$.

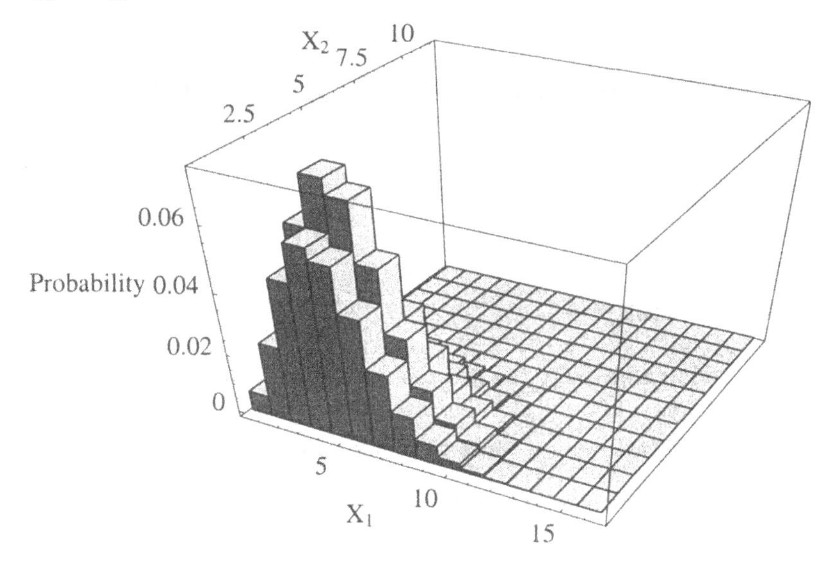

Expansion of the pde in (11.2) would give the set of Kolmogorov equations. The series expansion of (11.3) gives the following set of differential equations:

$$\begin{aligned}
\dot{\kappa}_{10}\ (t) &= k_{11}\kappa_{10} + k_{12}\kappa_{01} + I_1 \\
\dot{\kappa}_{01}\ (t) &= k_{21}\kappa_{10} + k_{22}\kappa_{01} + I_2 \\
\dot{\kappa}_{20}\ (t) &= k_{11}\kappa_{20} + k_{12}\kappa_{01} + 2k_{11}\kappa_{20} + 2k_{12}\kappa_{11} + I_1 \\
\dot{\kappa}_{11}\ (t) &= -k_{21}\kappa_{10} - k_{12}\kappa_{01} + k_{21}\kappa_{20} + (k_{11} + k_{22})\kappa_{11} + k_{12}\kappa_{02} \\
\dot{\kappa}_{02}\ (t) &= k_{21}\kappa_{10} + k_{22}\kappa_{01} + 2k_{21}\kappa_{11} + 2k_{22}\kappa_{02} + I_2
\end{aligned} \tag{11.12}$$

These equations generalize the single-population formulas in (4.20) and (4.21), and have solutions given in (11.11).

12

Linear Birth-Immigration-Death-Migration Models

12.1 Introduction

Consider now adding births to the previous linear multidimensional models discussed in Chapters 10 and 11. The assumptions are then

1. Prob$\{X_i$ will increase by 1 due to immigration$\}$ $= I_i\Delta t$

2. Prob$\{X_i$ will decrease by 1 due to death$\}$ $= \mu_i X_i \Delta t$

3. Prob $\{X_i$ will increase by 1 due to birth $\}$ $= \lambda_i X_i \Delta t$ (12.1)

4. Prob$\{X_i$ will increase by 1 and X_j will decrease by 1 due to migration $\}$ $= k_{ij}X_j\Delta t$ for $i, j = 1, 2$ with $i \neq j$.

These are the same as in Section 10.1 except for the additional birth assumption in (12.1).

The addition of births to the *deterministic* model does not entail profound changes in the nature of the solution. The previous deterministic model in Section 8.2 holds, with the previous death rate in compartment j, μ_j, replaced with the net death rate $(\mu_j - \lambda_j)$. Specifically, one could generalize the definition of k_{jj} in (8.2) to

$$k_{jj} = -\left(\mu_j - \lambda_j + \sum_{i=1, i\neq j}^{n} k_{ij}\right).$$

whereupon the general matrix model in (8.4) with solution in (8.5) and (8.6) holds. Moreover, if all the net deathrates are nonnegative, then one has negative eigenvalues and a classical compartmental analysis may be utilized.

As a stark contrast, however, the new *stochastic* models with births have solutions which are very different qualitatively from those without births in Chapters 10 and 11, as one might expect from the results in Chapter 5. Chiang [12] considers the special case of this model with births but without immigration. His application is to health-related demographics, and he calls it the birth-illness-death model. He observes that the "differential equations (for the probabilities) are unmanageable," hence "with the mathematics we now possess, we cannot obtain explicit formulas either for the probability generating function or the probabilities" [12, p. 497]. Therefore, though the system of Kolmogorov equations could be easily obtained, one is limited to a numerical solution, and in most cases of interest only for small t. Hence the Kolmogorov equations will not be considered in this chapter.

12.2 Equations for Cumulant Functions

Consider again for present simplicity a stochastic model with $n = 2$ compartments. The principles for any n are the same, and an application to an $n = 3$ compartment system will be discussed subsequently in Section 12.4. The equations in (12.1) for the $n = 2$ case lead to the following intensity functions:

$$\begin{aligned} f_{1,0} &= I_1 + \lambda_1 X_1 \qquad & f_{0,-1} &= \mu_2 X_2 \\ f_{0,1} &= I_2 + \lambda_2 X_2 \qquad & f_{-1,1} &= k_{21} X_1 \\ f_{-1,0} &= \mu_1 X_1 \qquad & f_{1,-1} &= k_{12} X_2 \end{aligned} \tag{12.2}$$

Substituting these into the operator equation for the mgf in (9.34), and transforming to the cgf, one has partial differential equations:

$$\begin{aligned} \frac{\partial K(\boldsymbol{\theta};t)}{\partial t} &= \left[\left(e^{\theta_1}-1\right)\lambda_1 + \left(e^{-\theta_1}-1\right)\mu_1 + \left(e^{-\theta_1+\theta_2}-1\right)k_{21}\right]\frac{\partial K(\boldsymbol{\theta};t)}{\partial \theta_1} \\ &+ \left[\left(e^{\theta_2}-1\right)\lambda_2 + \left(e^{-\theta_2}-1\right)\mu_2 + \left(e^{\theta_1-\theta_2}-1\right)k_{12}\right]\frac{\partial K(\boldsymbol{\theta};t)}{\partial \theta_2} \\ &+ \sum_{i=1}^{2}\left(e^{\theta_i}-1\right)I_i \end{aligned} \tag{12.3}$$

It is this equation which according to [12] does not have an explicit solution, even in the case $I_i = 0$. Therefore, we consider its direct expansion into equations for the cumulants. Let

$\mathbf{K}_m(t) = [\kappa_{m,0}(t), \kappa_{m-1,1}(t), \ldots, \kappa_{0,m}(t)]'$ be the vector of mth order cumulants, with

$\dot{\mathbf{K}}_m(t)$= derivative of $\mathbf{K}_m(t)$.

For example,

$\mathbf{K}_1(t) = [\kappa_{10}(t), \kappa_{01}(t)]'$ =vector of means

$\mathbf{K}_2(t) = [\kappa_{20}(t), \kappa_{11}(t), \kappa_{02}(t)]'$ = vector of variances and covariance.

Using the series expansion of the linear pde in (12.3), one can show that the differential equations for the cumulant functions have linear form:

$$\dot{\mathbf{K}}_m(t) = \sum_{i=1}^{m} \mathbf{A}_{mi}\mathbf{K}_i(t) + \mathbf{C}_m. \tag{12.4}$$

where the $\mathbf{A}_{ij}$ are pattern matrices and the $\mathbf{C}_j$ are input vectors. For example, the block diagonal $\mathbf{A}_{mm}$ matrices have form:

$$\mathbf{A}_{mm} = \begin{bmatrix} ma & mb & 0 & .. & 0 & 0 & 0 \\ c & (m-1)a+d & (m-1)b & .. & 0 & 0 & 0 \\ 0 & 2c & (m-2)a+2d & .. & 0 & 0 & 0 \\ \vdots & \vdots & \vdots & .. & \vdots & \vdots & \vdots \\ 0 & 0 & 0 & .. & (m-1)c & a+(m-1)d & b \\ 0 & 0 & 0 & .. & 0 & mc & md \end{bmatrix} \tag{12.5}$$

where

$$a = \lambda_1 - \mu_1 - k_{21}, \; b = k_{12}, \quad c = k_{21}, \text{ and } d = \lambda_2 - \mu_2 - k_{12}. \tag{12.6}$$

Also, one has

$$\mathbf{A}_{21} = \begin{bmatrix} (\lambda_1 + \mu_1 + k_{21}) & b \\ -c & -b \\ c & (\lambda_2 + \mu_2 + k_{12}) \end{bmatrix}. \tag{12.7}$$

Note, for example, that one could set up a system of differential equations for the five first and second order cumulants using (12.4)–(12.7). The explicit solutions to these cumulant functions is given in [82]. We will obtain numerical solutions to these and higher order cumulants in the illustrations to follow.

12.3 Application to Dispersal of African Bees.–Basic Model

The dispersal of the African bee using detailed data from French Guiana obtained in [75, 76] was investigated in [62]. The study envisioned two areas with trap lines 100 km. apart. The objective was to study the population growth in the two areas over an 8 month period due to a single initial colony in area 1. Hence the initial values were $X_1(0) = 1$ and $X_2(0) = 0$. Based on the empirical data in [75], the parameters for the model are:

$$I_1 = 0.050/\text{mo}, \qquad I_2 = 0 \tag{12.8}$$

$$\mu_1 = \mu_2 = 0.025/\text{mo}$$

$$\lambda_1 = \lambda_2 = 0.3055/\text{mo} \quad \text{and}$$

$$k_{12} = k_{21} = 0.0795/\text{mo.}$$

Upon substitution into (12.5) and (12.7), and also obtaining $\mathbf{A}_{31}$ and $\mathbf{A}_{32}$, the coefficient matrix for the system of nine cumulant functions of order 3 or less is

$$\mathbf{A} =$$

$$\begin{bmatrix} .2010 & .0795 & 0 & 0 & 0 & 0 & 0 & 0 & 0 \\ .0795 & .2010 & 0 & 0 & 0 & 0 & 0 & 0 & 0 \\ .4100 & .0795 & .4020 & .1590 & 0 & 0 & 0 & 0 & 0 \\ -.0795 & -.0795 & .0795 & .4020 & .0795 & 0 & 0 & 0 & 0 \\ .0795 & .4100 & 0 & .1590 & .4020 & 0 & 0 & 0 & 0 \\ .2010 & .0795 & 1.230 & .2385 & 0 & .6030 & .2385 & 0 & 0 \\ .0795 & -.0795 & -.1590 & .2510 & .0795 & .0795 & .6030 & .1590 & 0 \\ -.0795 & .0795 & .0795 & .2510 & -.1590 & 0 & .1590 & .6030 & .0795 \\ .0795 & .2010 & 0 & .2385 & 1.230 & 0 & 0 & .2385 & .6030 \end{bmatrix}$$

The solutions for the first and second order cumulant functions are given in Figures 12.1 and 12.2. The cumulant values at the end of the 8 month period are

$$\begin{aligned} &\kappa_{10} = 7.127 \quad \kappa_{01} = 3.807 \\ &\kappa_{20} = 55.27 \quad \kappa_{11} = 15.20 \quad \kappa_{02} = 23.32 \\ &\kappa_{30} = 788.3 \quad \kappa_{21} = 217.7 \quad \kappa_{12} = 133.7 \quad \kappa_{03} = 264.6. \end{aligned} \tag{12.9}$$

The bivariate distribution is intractable, as noted. Recently Renshaw [84] has developed bivariate saddlepoint approximations for this problem. For present simplicity, we consider finding only an approximating distribution by fitting these cumulants to those of some given parametric family. As an example, consider three variables, Y_1, Y_2 and Y_3, which follow negative binomial distributions, say $NB(k_1, p_1)$, $NB(k_2, p_2)$ and $NB(k_3, p_2)$. These negative binomials are constructed such that there are only five distinct parameters. Further, consider modeling at any specific time the population sizes X_1 and X_2 by the partial sums

$$\begin{aligned} X_1 &= Y_1 + Y_2, \text{ and} \\ X_2 &= Y_2 + Y_3, \end{aligned} \tag{12.10}$$

which, by construction, are obviously correlated. One could then find the five first and second order cumulants of the (X_1, X_2) vector as a function of the five parameters, as illustrated in [99]. Matching the cumulant expressions from (12.10) to the derived cumulants in (12.9), one would find parameter solutions

$$\begin{aligned} &k_1 = 0.609 \qquad k_2 = 0.484 \qquad k_3 = 0.259 \\ &p_1 = 0.116 \qquad p_2 = 0.163. \end{aligned}$$

Figure 12.1. Comparative mean value functions from bee population data for basic (ba), and Erlang $E(2)$ and $E(3)$ models, developed subsequently in the Appendix. Functions for swarming and multiple birth models are identical to basic model. Order (highest first) for means in comp. 1, κ_{10}, at $t = 8$ is ba, $E(2)$ and $E(3)$. Order for means in comp. 2, κ_{01}, at $t = 8$ is $E(3)$, $E(2)$, ba.

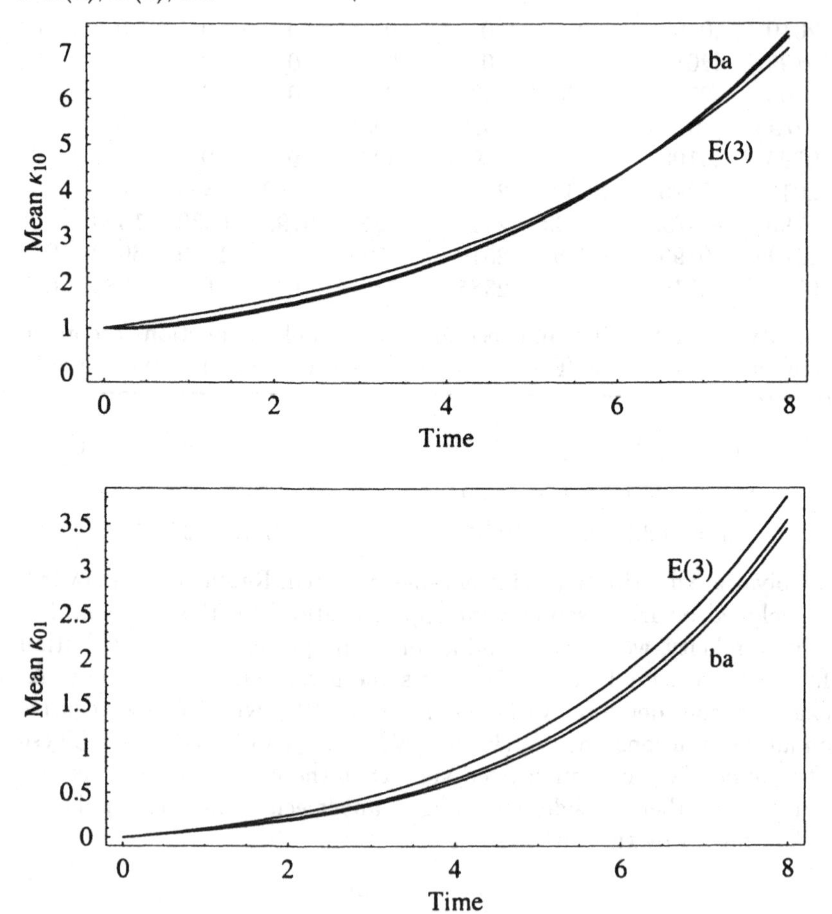

Figure 12.2. Comparative variance and covariance functions from bee population data for basic (ba), swarming (sw), multiple birth (mb), and Erlang E(2) and E(3) models. Models other than ba are developed in the Appendix.

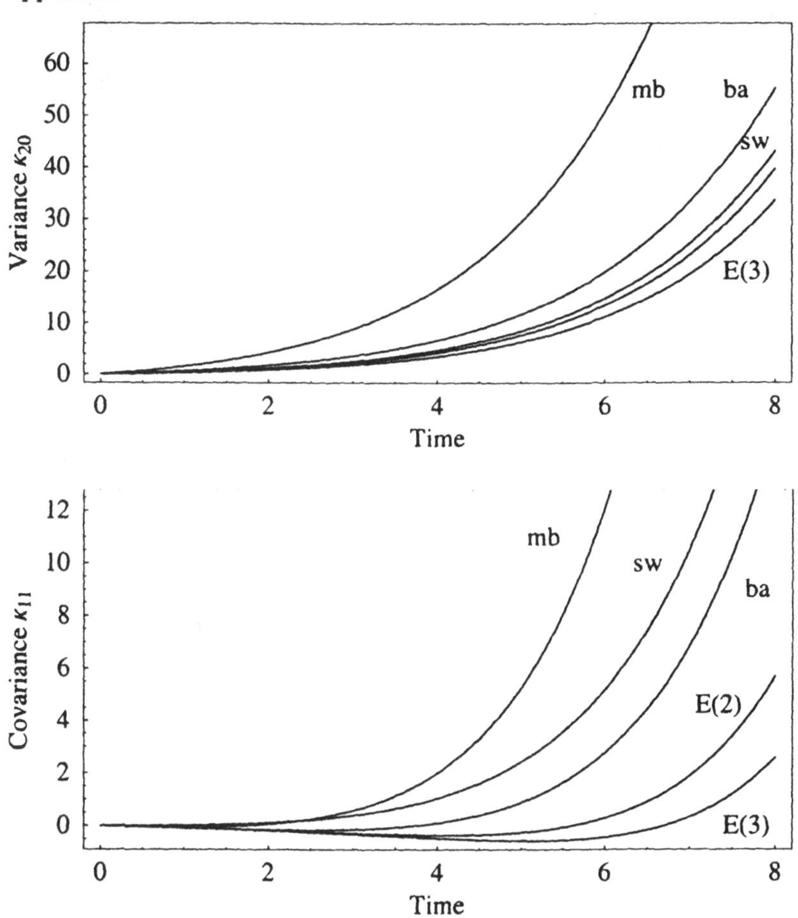

Figure 12.2. (Continued)

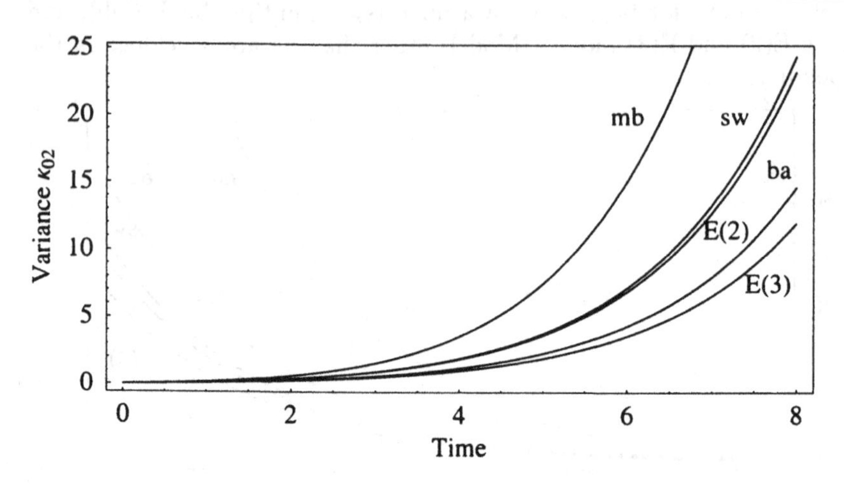

The resulting approximation at $t = 8$ using this given partial sums distribution is given in Figure 12.3, from [99]. By construction, the approximating bivariate distribution from (12.10) has the exact first and second order cumulants given in (12.9). The percent errors for the third order cumulants are 4.3, 21.4, 27.9 and 0.8% respectively.

Figure 12.3. Three-dimensional plot of approximating bivariate distribution from basic model for population sizes, X_1 and X_2, at $t = 8$.

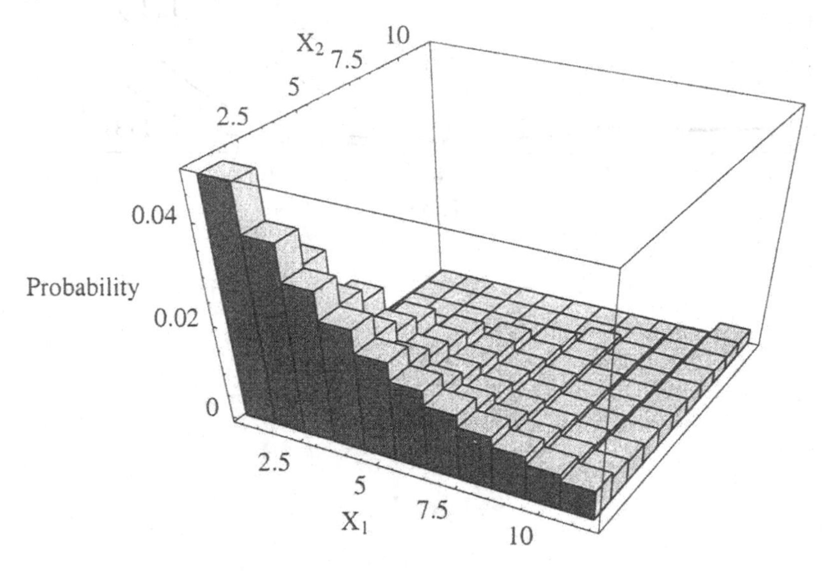

In summary, this "basic model" yields a seemingly reasonable bivariate distribution at any time over the 8 month period, using the assumed structure in (12.10) which is based on only five parameters. The standard approach has been to use computationally intensive daily simulation models, such as that in [75], to find the cumulant structure empirically. Models such as (12.10) represent an enormous simplification in the computational effort. Though the basic model fits the data well, one might question its biological realism. The model is critiqued from this standpoint in the Appendix, where also a number of alternative models are proposed and investigated.

12.4 Application to Muskrat Spread Data

Consider modeling also the muskrat spread data with the LBIDM model. Previously, the LBID model was fitted to selected adjacent individual provinces in Section 5.7.2. It has been standard practice to fit separate models to individual provinces [26], and the individual regression curves fit the data adequately for each province. However, the use of separate models for each province fails to take into account the migration effect between the provinces. Migration is an essential mechanistic component in many population management models, and hence we investigate the use of the LBIDM for these data.

For the purposes of this illustration, we consider the catch at three adjacent provinces, namely overijl, dr, and gron, on the eastern border of the Netherlands with Germany. We assume that the initial catch was zero in each province in 1969 (though overijl actually had an insignificant catch of 4 muskrats). Overijl is the northernmost and gron the southernmost of these three provinces, with dr in the middle. Hence geographical considerations rule out direct migration from overijl to gron. Also, the initial growth rate in dr lags those of the other provinces. Consequently, a simplifying assumption is made of one-way migration from overijl to dr and also from gron to dr. Also, another simplifying assumption is made that these three adjacent provinces have a common net birth rate, say $\lambda - \mu$. In principal, one can test such assumption statistically. Indeed, the assumption of equal net birth rates for a number of adjacent pairs and triples of provinces, is tested in [53] and the assumption is found to be tenable in each case.

The set of differential equations for the population sizes of the three provinces, for simplicity denoted as $X_1(t)$, $X_2(t)$, and $X_3(t)$, respectively, and with vector $\mathbf{X}(t) = [X_1(t), X_2(t), X_3(t)]'$ is

$$\dot{\mathbf{X}}(t) = \mathbf{A}_{11}\mathbf{X}(t) + \mathbf{C}_1 \tag{12.11}$$

where

$$\mathbf{A}_{11} = \begin{bmatrix} (\lambda-\mu)-k_{21} & 0 & \\ k_{21} & (\lambda-\mu) & k_{23} \\ 0 & 0 & (\lambda-\mu)-k_{23} \end{bmatrix} \tag{12.12}$$

and

$$\mathbf{C}_1 = [I_1, I_2, I_3]' \tag{12.13}$$

Note that $\mathbf{A}_{11}$ in (12.12) extends to three compartments the previous $\mathbf{A}_{11}$ matrix given in (12.5) for two compartments. This model has six identifiable parameters, namely $(\lambda-\mu)$, k_{21}, k_{23}, I_1, I_2 and I_3. The vector of mean value functions is defined as

$$\mathbf{K}_1(t) = [\kappa_{100}(t), \kappa_{010}(t), \kappa_{001}(t)]'. \tag{12.14}$$

The model for $\mathbf{K}_1(t)$, by virtue of this being a linear kinetic system, is

$$\dot{\mathbf{K}}_1(t) = \mathbf{A}_{11}\mathbf{K}_1(t) + \mathbf{C}_1 \tag{12.15}$$

as in (12.11).

As discussed in Section 5.7.2, the available data consists of observed catch, say $Y_i(t)$, in province i at elapsed time t. For simplicity, we again assume that the harvest rate, h, is time-invariant and is common to each province, i.e.

$$Y_i(t) = hX_i(t) \tag{12.16}$$

which extends the assumption (5.25) to multiple provinces. Let a cumulant function for the catch be denoted as $\kappa^\dagger(t)$, with

$$\mathbf{K}^\dagger(t) = [\kappa^\dagger_{100}(t), \kappa^\dagger_{010}(t), \kappa^\dagger_{001}(t)]. \tag{12.17}$$

It follows that the model for these transformed mean functions is

$$\dot{\mathbf{K}}^\dagger_1(t) = \mathbf{A}_{11}\mathbf{K}^\dagger_1(t) + h\mathbf{C}_1, \tag{12.18}$$

i.e. the net birthrate and migration rates are invariant to the transformation, and the new immigration rates are $I^\dagger_i = hI_i$. The generalization to different but time-invariant rates among the provinces is discussed in [53].

The parameter estimates, with their standard errors, from fitting model (12.18) to the data are as follows. The estimated common net birthrate $(\lambda-\mu)$ is 0.328 (0.023). The estimated migration rates are $k_{21} = 0.032(0.033)$ and $k_{23} = 0.071(0.024)$. The estimated net rates of increase for overjil and gron are therefore 0.328-0.032 = 0.296 and 0.328-0.071 = 0.258, which are identical to those estimated for these individual provinces in Section 5.7.2. However, the new estimated net birthrate for dr is 0.328, as opposed to the estimate of 0.421 when this province was fitted individually. The estimated immigration rates, $I^\dagger_i$, are 174 (48), 61 (73) and 343(60). Again, the estimates for overijl and gron are the same as those obtained previously, however the new estimate of 61 for dr is considerably smaller than the previous estimate of 115 from the individual model. Clearly, the residual sum of squares for the overijl and gron data are the same as before, however this measure of model fit reduces to RSS=3.48 for the combined model as compared to the previous RSS = 4.00.

When the three sets of data were fitted *individually*, it was found that the estimated net birth rate for dr is significantly ($p < .05$) larger than the net rates for the adjacent provinces. In a simultaneous fitting, however, there is no reason to believe that the net birthrates are different, but rather that there is net migration into dr from both overijl and gron. Similar qualitative findings, which not only estimate the net migrations of interest but also provide ecologically reasonable interpretations, were found for many of the other provinces in this rich data set.

12.5 Appendix

12.5.1 Application to Dispersal of African Bees–Generalized Model

12.5.1.1 Critique of Basic Model for Dispersal of African Bees

The previous model provides an adequate fitting to the very limited available data. It is also a very useful management tool, as it is based on kinetic parameters and is computationally relatively easy. However, the model also has the following tacit assumptions, which are *not* biologically realistic:

i) constant birth, death, and migration rates over time,

ii) a so-called "simple" birth mechanism in which each birth probability is a linear function only of its respective population size,

iii) unit births, i.e. clutch or litter size of 1,

iv) exponentially distributed time intervals between migrations and between birth events.

Models which relax each of these assumptions individually follow, as well as a generalized model which relaxes the latter three together.

12.5.1.2 Model with Time-Varying Rates

The first assumption could be relaxed by using piece-wise constant approximations for time varying rates. This is illustrated in [54], where three different periods, each with a different parameter vector, are used.

12.5.1.3 Model with Swarming

When bee colony reproduction called swarming occurs, a new colony at the time of its "birth" may locate in some other area. Hence, the birth rate in an area may be a linear function of the population sizes in all of the areas. Consider generalizing the intensity functions describing birth in (12.2) to

$$\begin{aligned} f_{1,0} &= I_1 + \lambda_{11}X_1 + \lambda_{12}X_2 \\ f_{0,1} &= I_2 + \lambda_{22}X_2 + \lambda_{21}X_1, \end{aligned} \tag{12.11}$$

where λ_{12} and λ_{21} are called the cross-compartment birth rates.

The new pde for the cgf is:

$$\begin{aligned} \frac{\partial K(\boldsymbol{\theta};t)}{\partial t} &= \Big[\left(e^{\theta_1}-1\right)\lambda_{11} + \left(e^{\theta_2}-1\right)\lambda_{21} \\ &\quad + \left(e^{-\theta_1}-1\right)\mu_1 + \left(e^{-\theta_1+\theta_2}-1\right)k_{21} \Big] \frac{\partial K(\boldsymbol{\theta};t)}{\partial \theta_1} \\ &\quad + \Big[\left(e^{\theta_1}-1\right)\lambda_{12} + \left(e^{\theta_2}-1\right)\lambda_{22} \\ &\quad + \left(e^{-\theta_2}-1\right)\mu_2 + \left(e^{\theta_1-\theta_2}-1\right)k_{12} \Big] \frac{\partial K(\boldsymbol{\theta};t)}{\partial \theta_2} \\ &\quad + \sum_{i=1}^{2}(e^{\theta_i}-1)I_i \end{aligned} \tag{12.12}$$

The cumulant equations retain the linear form in (12.4), though the structure of the $\mathbf{A}_{ij}$ matrices changes, of course.

In the current context, it is estimated that a swarm moves to the other compartment with probability $s = 0.25$. The model was reparameterized to produce the same mean value functions, $\kappa_{10}(t)$ and $\kappa_{01}(t)$, as before with the basic model in (12.8). Specifically, the new parameter values were

$$\begin{aligned} \lambda_{11} &= \lambda_{22} = 0.2291 \\ \lambda_{12} &= \lambda_{21} = 0.0794 \\ k_{12} &= k_{21} = 0.0031. \end{aligned}$$

The second-order cumulants for this model with swarming are plotted in Figure 12.2. Note that the main difference is the increase in the covariance in the swarming model, as one would expect.

12.5.1.4 Model with Multiple Births

Consider changing the birth rate assumption in (12.1) to incorporate multiple births. Specifically let:

$$\text{Prob}\{X_i \text{ will increase by } k \text{ due to birth }\} = \lambda_i p_i(k) X_i \Delta t, \tag{12.13}$$

where the $p_i(k)$, for $k = 0, 1, \ldots, N < \infty$; are probabilities such that $\sum_k p_i(k) = 1$ for all i.

The corresponding intensity functions for births in (12.2) change to:

$$\begin{aligned} f_{1,0} &= I_1 + \lambda_1 p_1(1) X_1, \\ f_{0,1} &= I_2 + \lambda_2 p_2(1) X_2, \end{aligned} \tag{12.14}$$

$$f_{k_1,0} = \lambda_1 p_1(k_1) X_1, \qquad \text{for } k_1 > 1, \quad \text{and}$$
$$f_{0,k_2} = \lambda_2 p_2(k_2) X_2, \qquad \text{for } k_2 > 1.$$

These give the pde for the cgf as:

$$\frac{\partial K(\boldsymbol{\theta};t)}{\partial t} =$$
$$\left[\lambda_1 \sum_{k\geq 1}(e^{k\theta_1}-1)p_i(k) + \left(e^{-\theta_1}-1\right)\mu_1 + \left(e^{-\theta_1+\theta_2}-1\right)k_{21}\right]\frac{\partial K(\boldsymbol{\theta};t)}{\partial \theta_1}$$
$$+\left[\lambda_2 \sum_{k\geq 1}(e^{k\theta_2}-1)p_2(k) + \left(e^{-\theta_2}-1\right)\mu_2 + \left(e^{\theta_1-\theta_2}-1\right)k_{12}\right]\frac{\partial K(\boldsymbol{\theta};t)}{\partial \theta_2}$$
$$+\sum_{i=1}^{2}\left(e^{\theta_i}-1\right)I_i \tag{12.15}$$

For example, in the present modeling problem, the empirical distribution from [75, 76] for the number of swarms observed per swarming episode in French Guiana is

k	1	2	3	4	5
$p(k)$	0.17	0.28	0.35	0.14	0.06

(12.16)

which yields a mean of 2.64 and a standard deviation of 1.10 swarms per swarming episode (i.e. birth event). These data were incorporated into (12.14) and (12.15).

The cumulant equations for the multiple birth model again have the linear form in (12.4). This model also was reparameterized to yield the same mean value functions as in the basic model. Conceptually, the principal change was to slow down the birth episodes (by a factor of 2.64) to compensate for the increase in the mean number of births, from the previous mean of 1.0 to the new mean of 2.64. The second-order cumulant functions from this model are also plotted in Figure 12.2. It is apparent that all of the second-order cumulants increase substantially as compared to the basic model, reflecting this large additional source of uncertainty.

12.5.1.5 Model with Erlang Distributed Waiting Times

The previous generalizations retained the assumption of exponential waiting times between birth episodes. Consider generalizing the birth assumption in (12.1) to

Prob{ the X_i colonies of age a would increase by 1 due to birth from a to

$$a + \Delta a\} = \lambda_i(a)X_i\Delta a + o(\Delta a), \tag{12.17}$$

which is analogous to the age-dependent hazard rates in (10.32). Previous study of Otis' data has indicated that the assumption of Erlang distributed waiting times between births is a reasonable approximation for the data. A schematic for the LBIDM model representation with assumed Erlang $(2, \lambda_1)$ and Erlang $(2, \lambda_2)$ waiting time distributions in the two areas, together with the previously assumed immigration, death and migration rates in (12.1), is given in Figure 12.4. The generalization to Erlang shape parameters n_1 and n_2 is immediate; in effect, the previous model with two populations with nonexponential times would be represented by $(n_1 + n_2)$ populations with exponential times, as illustrated in [62].

Figure 12.4. Pseudo-compartment representation of a non-Markovian dispersal model with Erlang (2) and Erlang (2) distributions of time intervals between births.

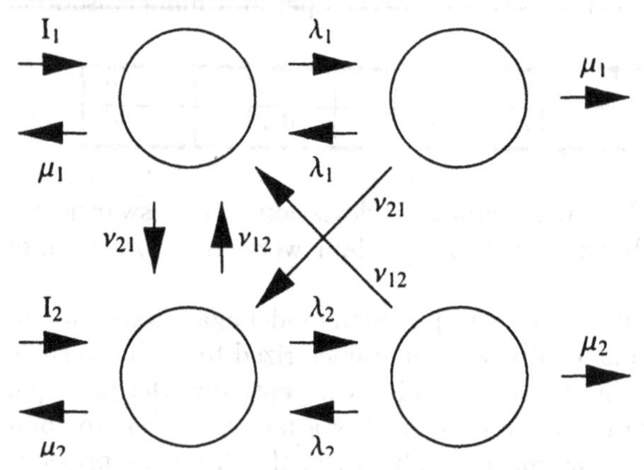

Consider the following notation for the counts, cumulants and intensity functions for the new pseudo-compartments. Let

1. $X_{ij}(t)$, for $i = 1, 2; j = 1, 2, \ldots, n_i$; denote the random count of units in pseudo-compartment j of original compartment i;

2. $\kappa^*_{v_{11},v_{12},\ldots,v_{2n2}}(t)$ denote a joint cumulant of the random vector $\mathbf{X}^*(t) = [X_{11}(t), \ldots, X_{1,n1}(t), X_{21}(t), \ldots, X_{2,n2}(t)]$.

3. Prob{ $X_{ij}(t)$ changes by u_{ij} units, for all i and j } = $f_{u_{11},u_{12},\ldots,u_{2,n2}}[X_{11}, X_{12}, \ldots, X_{2,n2}]\Delta t$

As an example, the intensity functions for the immigration, death, and migration events for the special case of $n_1 = n_2 = 2$ are

$$f_{1,0,0,0} = I_1 \qquad f_{-1,0,1,0} = k_{21}X_{11}$$
$$f_{0,0,1,0} = I_2 \qquad f_{0,-1,1,0} = k_{21}X_{12}$$

$$\begin{aligned} f_{-1,0,0,0} &= \mu_1 X_{11} & f_{1,0-1,0} &= k_{12} X_{21} \\ f_{0,-1,0,0} &= \mu_1 X_{12} & f_{0,1,0,-1} &= k_{12} X_{22} \\ f_{0,0,-1,0} &= \mu_2 X_{21} \\ f_{0,0,0,-1,} &= \mu_2 X_{21} \end{aligned} \tag{12.18}$$

The corresponding functions for generating the desired births at $E(2)$ distributed intervals are:

$$\begin{aligned} f_{-1,1,0,0} &= \lambda_1 X_{11} & f_{0,0,-1,1} &= \lambda_2 X_{21} \\ f_{2,-1,0,0} &= \lambda_1 X_{12} & f_{0,0,2,-1} &= \lambda_2 X_{22}. \end{aligned} \tag{12.19}$$

The partial differential equation for the joint distribution of $\mathbf{X}^*(t)$ is

$$\begin{aligned} \frac{\partial K}{\partial t} &= \\ &\sum_{i=1}^{n_1-1} \left[\mu_1(e^{-\theta_{1i}} - 1) + k_{21}(e^{-\theta_{1i}+\theta_{21}} - 1) + \lambda_1(e^{-\theta_{1i}+\theta_{1,i+1}} - 1)\right] \frac{\partial K}{\partial \theta_{1i}} \\ &+ \left[\mu_1(e^{-\theta_{1,n1}} - 1) + k_{21}(e^{-\theta_{1,n1}+\theta_{21}} - 1) + \lambda_1(e^{-\theta_{1,n1}+2\theta_{11}} - 1)\right] \frac{\partial K}{\partial \theta_{1,n1}} \\ &+ \sum_{i=1}^{n_2-1} \left[\mu_2(e^{-\theta_{2i}} - 1) + k_{12}(e^{-\theta_{2i}+\theta_{11}} - 1) + \lambda_2(e^{-\theta_{2i}+\theta_{2,i+1}} - 1)\right] \frac{\partial K}{\partial \theta_{2i}} \\ &+ \left[\mu_2(e^{-\theta_{2,n2}} - 1) + k_{12}(e^{-\theta_{2,n2}+\theta_{11}} - 1) + \lambda_2(e^{-\theta_{2,n2}+2\theta_{21}} - 1)\right] \frac{\partial K}{\partial \theta_{2,n2}} \\ &+ \sum_{i=1}^{2} I_i(e^{\theta_{i1}} - 1) \end{aligned} \tag{12.20}$$

After solving for the joint cumulants, $\mathbf{K}^*$, for the expanded system, one may transform them back to cumulants for the original Erlang system. For example, for $n_1 = n_2 = 2$, using the relationships

$$X_i(t) = \sum_{j=1}^{n_1} X_{ij}(t),$$

the cumulant transformations are:

$$\begin{aligned} \kappa_{10} &= \kappa^*_{1,0,0,0} + \kappa^*_{0,1,0,0} \\ \kappa_{01} &= \kappa^*_{0,0,1,0} + \kappa^*_{0,0,0,1} \\ \kappa_{20} &= \kappa^*_{2,0,0,0} + 2\kappa^*_{1,1,0,0} + \kappa^*_{0,2,0,0} \\ \kappa_{11} &= \kappa^*_{1,0,1,0} + \kappa^*_{1,0,0,1} + \kappa^*_{0,1,1,0} + \kappa^*_{0,1,0,1} \\ \kappa_{02} &= \kappa^*_{0,0,2,0} + 2\kappa^*_{0,0,1,1} + \kappa^*_{0,0,0,2} \end{aligned} \tag{12.21}$$

In illustrating this model, it is not possible to reparameterize the Erlang model to give the identical mean value functions, $\kappa_{10}(t)$ and $\kappa_{01}(t)$, as previously found for the basic model with exponential waiting times. However, close approximations may be obtained. Figure 12.1 illustrates the comparable mean value functions for the two cases where both areas have identical Erlang (2) distributions and also where both have identical Erlang (3) distributions. The new mean functions are close to those for the exponential case, but it is clear in Figure 12.2 that the Erlang models have smaller second-order cumulant functions.

12.5.1.6 Model with Swarming, Multiple Births and Erlang Distributed Waiting Times

Consider now a model which combines the last three generalizations. We assume k births per swarming episode, with probabilities in (12.16), where each independent swarm originating in area i has probability s_i of staying in i and probability $(1 - s_i)$ of moving, and where the waiting times are Erlang (2) distributed in each area. The intensity functions for j_1 births in 1 and j_2 in 2, such that $k = j_1 + j_2$, are

$$f_{j_1+1,-1,j_2,0} = \lambda_1 \binom{j_1 + j_2}{j_1} s_1^{j_1}(1 - s_1)^{j_2} p[j_1 + j_2] X_{12}, \text{ and}$$

$$f_{j_1+1,0,j_2+1,-1} = \lambda_2 \binom{j_1 + j_2}{j_2} (1 - s_2)^{j_1} s_2^{j_2} p[j_1 + j_2] X_{22},$$

for $j_1 = 1, 2, \ldots, k$ and $j_2 = 1, 2, \ldots, k - j_1$.

The pde which generalizes (12.3) is

$$\frac{\partial K}{\partial t} =$$

$$\sum_{i=1}^{n_1-1} \left[\mu_1(e^{-\theta_{1i}} - 1) + k_{21}(e^{-\theta_{1i}+\theta_{21}} - 1) + \lambda_1(e^{-\theta_{1i}+\theta_{1,i+1}} - 1)\right] \frac{\partial K}{\partial \theta_{1i}}$$

$$+ \left[\mu_1(e^{-\theta_{1,n1}} - 1) + k_{21}(e^{-\theta_{1,n1}+\theta_{21}} - 1) + \lambda_1 \sum_{j_1=0}^{k} \sum_{j_2=0}^{k-j_1} \binom{j_1 + j_2}{j_1} \cdot \right.$$

$$\left. s_1^{j_1}(1 - s_1)^{j_2} p_1[j_1 + j_2](e^{(j_1+1)\theta_{11} + j_2\theta_{21} - \theta_{1,n1}})\right] \frac{\partial K}{\partial \theta_{1,n1}}$$

$$+ \sum_{i=1}^{n_2-1} \left[\mu_2(e^{-\theta_{2i}} - 1) + k_{12}(e^{-\theta_{2i}+\theta_{11}} - 1) + \lambda_2(e^{-\theta_{2i}+\theta_{2,i+1}} - 1)\right] \frac{\partial K}{\partial \theta_{2i}}$$

$$+ \left[\mu_2(e^{-\theta_{2,n2}} - 1) + k_{12}(e^{-\theta_{2,n2}+\theta_{11}} - 1) + \lambda_2 \sum_{j_1=0}^{k} \sum_{j_2=0}^{k-j_1} \binom{j_1 + j_2}{j_2} \cdot \right.$$

$$\left. (1 - s_1)^{j_1} s_2^{j_2} p_2[j_1 + j_2](e^{j_1\theta_{11} + (j_2+1)\theta_{21} - \theta_{2,n2}})\right] \frac{\partial K}{\partial \theta_{1,n2}}$$

$$+\sum_{i=1}^{2}(e^{\theta_{i1}}-1)I_i \tag{12.22}$$

The model was parameterized using the observed data for multiple births, the conjectured cross-compartment birth rates, and the best fitting Erlang (2) through Erlang (7) waiting time distributions. With Erlang (7) waiting times, there are 14 pseudo-compartments, which have a total of 679 differential equations for the first, second and third order cumulants. The results for each of these models, when transformed back to original first and second order cumulants, are illustrated in Figure 12.5 and 12.6. The model with Erlang (5) waiting times yielded the best fitting to the observed variance in [75, 76].

The cumulants for the Erlang (5) model at $t = 8$ are

$$\begin{aligned}
&\kappa_{10} = 7.64 \quad \kappa_{01} = 3.29 \\
&\kappa_{20} = 40.17 \quad \kappa_{11} = 15.32 \quad \kappa_{02} = 15.91 \\
&\kappa_{30} = 388.9 \quad \kappa_{21} = 148.3 \quad \kappa_{12} = 107.2 \quad \kappa_{03} = 138.9.
\end{aligned} \tag{12.23}$$

Using the previous partial sums of negative binomials model in (12.10), the parameter solutions for the new approximating distribution are

$$\begin{aligned}
&k_1 = 0982 \qquad k_2 = 0.825 \qquad k_3 = 0.0318 \\
&p_1 = 0.180 \qquad p_2 = 0.207
\end{aligned}$$

This bivariate distribution is illustrated in Figure 12.7. This distribution has the exact first and second order cumulants in (12.23), with errors in the third order cumulants of 1.2, 10.4, 23.9 and 0.6% respectively.

Clearly, as one compares the basic model in Section 12.3 to this generalized model, it is obvious that the former is much easier to conceptualize and compute. However, the present example demonstrates that mathematical tools exist to make the model more realistic biologically. From a quantitative perspective, differences in the two bivariate approximations in Figures 12.3 and 12.7 might not be immediately visible. Nevertheless, there are substantial differences in the shape characteristics of the marginal distributions, which are given in [99] but not displayed here. From a theoretical standpoint, the generalized model which incorporates greater biological realism and mathematical structure is obviously the preferred model for making predictions; yet from a practical standpoint, the much simpler basic model may be adequate for the given objectives.

Figure 12.5. Comparative mean value functions from bee population data for basic (ba) and for full Erlang $E(2)$ through $E(5)$ models. Order (highest first) for means in comp. 1, κ_{10}, is $E(5)$, $E(4)$, $E(3)$, $E(2)$, and ba. Order for means in comp. 2, κ_{01}, is ba, $E(2)$, $E(3)$, $E(4)$ and $E(5)$.

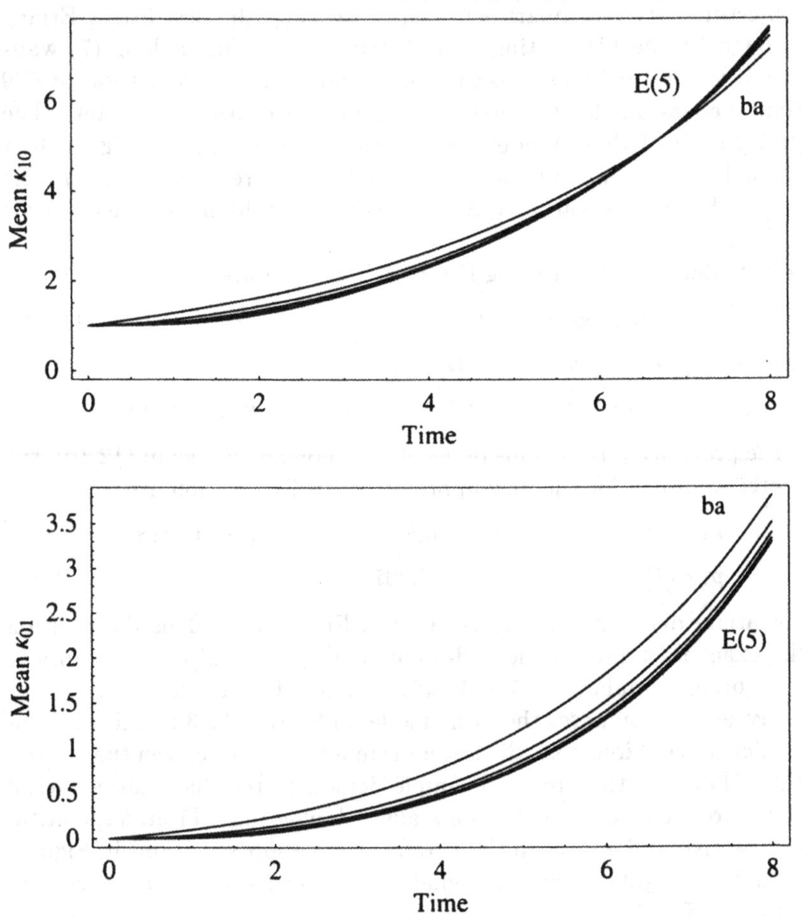

Figure 12.6. Comparative variance and covariance functions from bee population date for basic (ba) and for full Erlang $E(2)$ through $E(5)$ models.

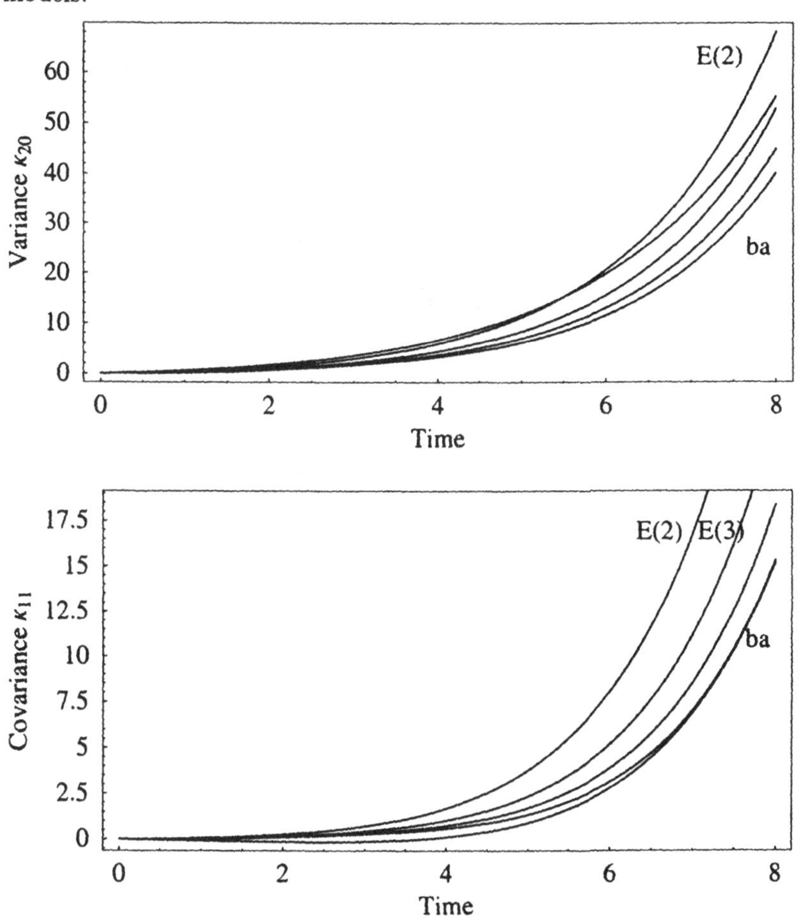

Figure 12.6. (Continued)

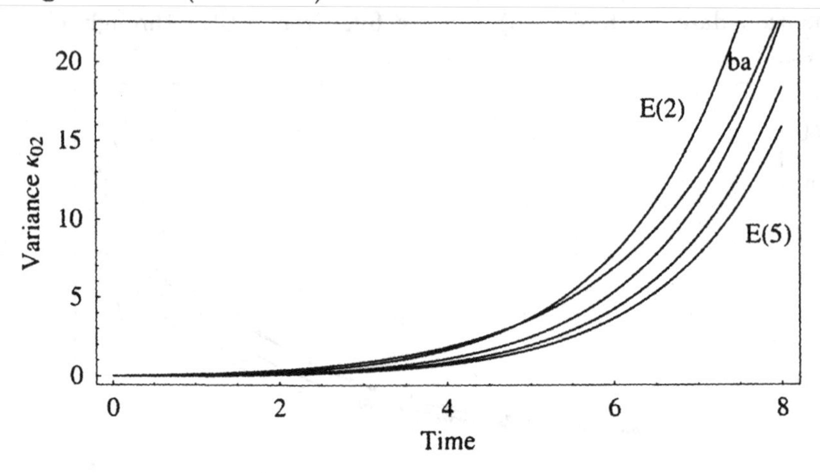

Figure 12.7. Three-dimensional plot of the approximating bivariate distribution from the full Erlang E(5) model for population sizes, X_1 and X_2, at $t = 8$.

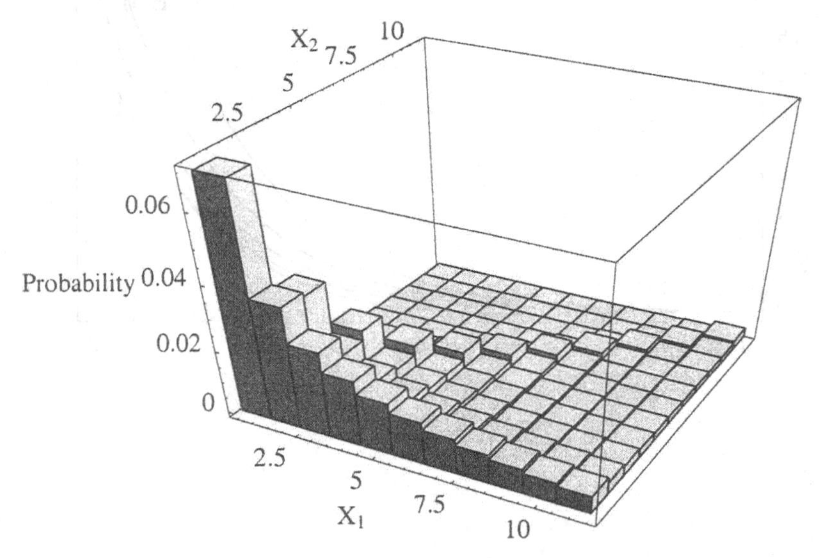

13

Nonlinear Birth-Death-Migration Models

13.1 Introduction

Consider generalizing the single-population models with nonlinear kinetics given in Chapters 6 and 7 to the multiple population case. Chapters 13 and 14 consider two general models, very different from one another, to illustrate the application of the methodology introduced in Chapter 9 to various multi-population models with nonlinear rates.

We consider first populations linked by migration. The LBIDM model in Chapter 12 based on linear kinetics often provides an adequate basis for describing population growth and dispersal for a single species when population sizes are relatively small. The model also has a stable equilibrium distribution, but only in the case where the death rates, μ_i, exceed the corresponding birthrates, λ_i. This case may not be realistic for many populations; specifically it is not realistic for long-term AHB population dynamics.

In order to generate transient multivariate distributions which lead to a credible equilibrium distribution as populations grow over time, consider modifying assumptions (12.1) in Chapter 12 to include the density-dependent birth and death rates, as given in (6.1) and (6.2) (and also in (7.1) and (7.2)). The motivation from a practical standpoint for the present model is to investigate the effect of simple migration on the ensuing transient and equilibrium distributions. With this objective in mind, two simpifications are made. The immigration effect is not included in the present model, for the sake of parameter parsimony. Also, the model assumes only unidirectional (one-way) migration, in order to quantify more readily the effect of net migration.

For present expediency, only the case of two compartments will be developed. Formally, the assumptions of unit changes, in a small time increment from t to $t + \Delta t$, for the present model of interest are:

1. Prob$\{X_i$ will increase by 1 due to birth$\} = \lambda_{i,X_i} \Delta t$

$$\text{where } \lambda_{i,X_i} = \begin{cases} a_{i1}X_i - b_{i1}X_i^{s+1} & \text{for } X_i < (a_{i1}/b_{i1})^{1/s} \\ 0 & \text{otherwise} \end{cases} \tag{13.1}$$

2. Prob$\{X_i$ will decrease by 1 due to death$\} = \mu_{i,X_i}\Delta t$

$$\text{where } \mu_{i,X_i} = a_{i2}X_i + b_{i2}X_i^{s+1} \tag{13.2}$$

3. Prob$\{X_2$ will increase by 1 and X_1 will decrease by 1 due to migration$\}$

$$= k_{21}X_1 \tag{13.3}$$

for $i = 1$ or 2, and for integer $s \geq 1$.

Figure 13.1 illustrates a schematic of the present model. Because of the one-way migration in the model, the first population, X_1, may be considered separately. Its marginal distribution may be obtained from the previous results in Chapter 6, with the birthrate given in (13.1) and the new effective deathrate consisting of the combined deathrate and migration rate, in (13.2) and (13.3), respectively. Clearly, however, the bivariate distribution is necessary in order to quantify the effect of migration in the model.

Figure 13.1. Schematic of NBDM model with one-way migration.

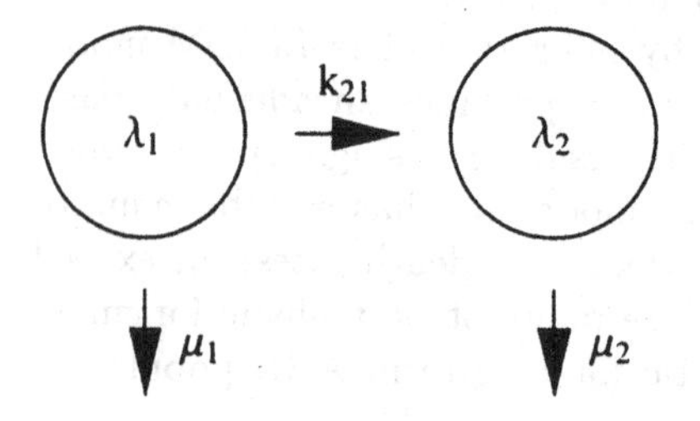

This restricted model, without immigration or two-way migration, is sufficient not only to investigate the immediate problem at hand, but also to illustrate briefly some of the underlying challenges in the multidimensional modeling of nonlinear kinetic systems. The resulting cumulant equations could be easily expanded to include both of these additional effects which are presently excluded.

13.2 Probability Distribution for the Stochastic Model

13.2.1 Derivation of Quasi-Equilibrium Distribution

One could write the system of Kolmogorov equations for the transient bivariate distributions. Both population sizes are clearly bounded, though the upper bound on the second population is a function of the rates of both populations. Approximating distributions may be obtained by truncating the equations at large upper limits, say U_1 and U_2. For the sake of simplicity, such bivariate solution of the Kolmogorov equations is not pursued in this chapter, however a bivariate solution is illustrated in Chapter 14.

Though the exact bivariate transient distributions for $X_1(t)$ and $X_2(t)$, as well as the transient moments of these distributions, are difficult to obtain in practice, the same is not true for the equilibrium distribution. Indeed, assuming that it exists, one can derive the exact quasi-equilibrium distribution for this uni-directional model relatively easily. The quasi-equilibrium distribution, in turn, demonstrates some of the effects of interest concerning migration.

Consider the following procedure which generalizes that in Section 6.3.2.

i. Let Prob$[N_1 = n_1, N_2 = n_2]$ be denoted $P(n_1, n_2)$. Using the fundamental conditional probability argument, one may write

$$P(n_1, n_2) = P(n_1)P(n_2|n_1) \tag{13.4}$$

where $P(n_1)$ is the marginal distribution of N_1 and $P(n_2|n_1)$ is the conditional distribution of N_2 given $N_1 = n_1$.

ii. To find $P(n_1)$, note that when the first population is in equilibrium, one has the recurrence relationship

$$\lambda_{1,n_1-1}P(n_1 - 1) = (\mu_{1,n_1} + n_1 k_{21})P(n_1),$$

with λ, μ and k_{21} defined in (13.1)–(13.3). This generalizes (6.13) but preserves the same form. Hence one can solve for $P(n_1)$ numerically, as before.

iii. To find the conditional distribution $P(n_2|n_1)$, note that in equilibrium, for given $N_1 = n_1$, one has the following recurrence relationship for N_2:

$$[\lambda_{2,n_2-1} + n_1 k_{21}]P(n_2 - 1) = \mu_{2,n_2}P(n_2) \tag{13.5}$$

In effect, (13.5) rules out a drift, because the probability of population size $(n_2 - 1)$ with a unit increase due to "birth" or "migration" equals the probability of size n_2 with a unit "death". One could solve (13.5) numerically to obtain $P(n_2|n_1)$ for all possible n_1 from 1 to $u_1 = (a_{11}/b_{11})^{1/s}$.

iv. The bivariate distribution is then immediate from (13.4).

Clearly ultimate extinction is certain also in this birth-death migration (NBDM) model, hence the result is only a quasi-equilibrium distribution, as in Section 6.3.2.

13.2.2 Application to AHB Migration Problem

As noted in Section 2.2, the movement of the leading edge of the AHB range expansion has slowed down considerably in southern U.S. [33]. The question of how a reduced migration rate would affect the ultimate equilibrium distribution is considered in [46] and in this section.

Consider a very restricted, though reasonable, case of this model, in which the two individual AHB populations linked by migration have identical "overall" birth and death rate functions. The populations are assumed to have the same crowding coefficients in rate functions (13.1)–(13.3), i.e. $b_{11} = b_{12} = b_1$ and $b_{12} = b_{22} = b_2$, and also the same intrinsic birth coefficient, i.e. $a_{11} = a_{21} = a_1$. Let the intrinsic death rate of the first population be set equal to the intrinsic rate of the second less the migration rate k_{21}. The model would be represented by Figure 13.1 with assumed rates for case $s = 1$ from (13.1)–(13.3) as follows:

$$\lambda_{1,X_1} = a_1 X_1 - b_1 X_1^2 \qquad \mu_{1,X_1} = (a_{22} - k_{21})X_1 + b_2 X_1^2$$
$$\lambda_{2,X_2} = a_1 X_2 - b_1 X_2^2 \qquad \mu_{2,X_2} = a_{22} X_2 + b_2 X_2^2$$

Consider first the bivariate distribution of X_1 and X_2 for this population model for the standard model with $s = 1$, first introduced in Section 6.2.2. The assumed rates are

$$a_1 = 0.30 \qquad a_{22} = 0.02$$
$$b_1 = 0.015 \qquad b_2 = 0.001.$$

The standard migration rate is set at $k_{21} = 0.015$, which is the best estimate of historical spread used in [62]. Therefore, the intrinsic deathrate parameters are

$$a_{12} = 0.005 \quad \text{and} \quad a_{22} = 0.02.$$

The marginal equilibrium distribution of X_1 for this standard case was previously given in Figure 6.3. The bivariate equilibrium distribution of X_1 and X_2 obtained using the methodology in Section 13.2.1 is illustrated in Figure 13.2.

The other specific migration levels investigated are $k_{21} = 0.020$ (which is the maximum possible for this model), 0.010, 0.005 and 0.001. The last three rates are added to study the impact of a slowdown in migration.

The cumulants for the X_1 distribution are given in Table 6.1 as $\kappa_{10} = 17.3564$, $\kappa_{20} = 2.4917$ and $\kappa_{30} = -2.2101$, which yield an index of skewness of $\gamma_1 = -0.562$. Table 13.1 gives for each of the five migration levels the mean, variance, and skewness, denoted κ_{01}, κ_{02}, and κ_{03}, for the marginal distribution of X_2, and the covariance κ_{11} between X_1 and X_2. It also gives the carrying capacities, K_2, for the corresponding deterministic model.

It is apparent in Table 13.1 that, as the migration rate k_{21} increases, the mean κ_{01} and the carrying capacity K_2 increase, as expected; and the covariance κ_{11} and correlation coefficient p also increase, as expected, reflecting the stronger expected dependency of X_2 on X_1. For this level of migration the variance κ_{02} increases, though the skewness, κ_{03}, and the index of skewness decrease in absolute value, implying a more symmetric distribution. The study is repeated in [46] for other values of a_1 for $s = 1$

Figure 13.2. Bivariate equilibrium distribution of the assumed NBDM model for AHB migration.

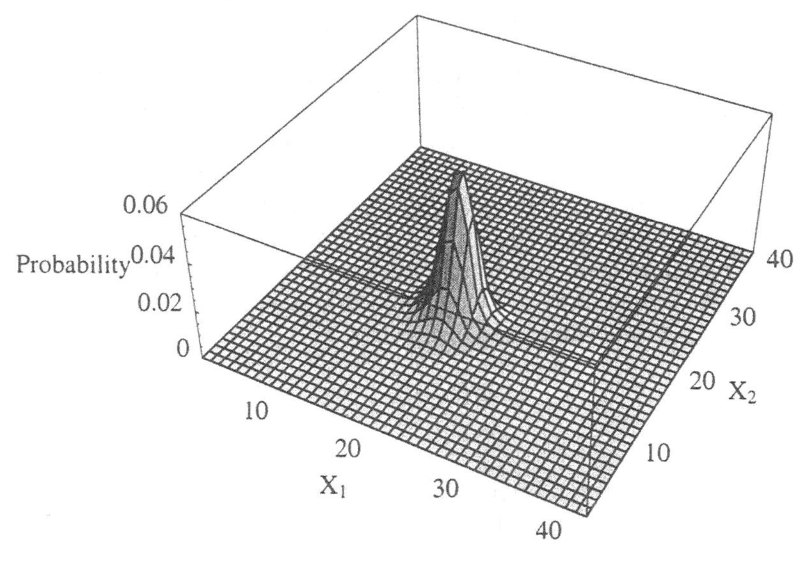

Table 13.1. Carrying capacity, K_2; cumulants associated with X_2, i.e. mean, κ_{01}, variance, κ_{02}, covariance, κ_{11}, and skewness, κ_{03}; correlation coefficient, ρ; and index of skewness, γ_2, for the bivariate model with $s = 1$ and five migration levels, k_{21}.

k_{21}	K_2	κ_{01}	κ_{02}	κ_{11}	ρ	κ_{03}	γ_2
.02	18.6716	18.714	3.005	0.218	0.080	-0.329	-0.063
.015	18.3920	18.347	2.728	0.151	0.058	-1.225	-0.272
.01	18.10141	18.006	2.574	0.095	0.038	-1.756	-0.425
.005	17.8071	17.679	2.504	0.046	0.019	-2.054	-0.518
.001	17.5623	17.421	2.491	0.009	0.004	-2.187	-0.556

and $s = 2$. In all these other cases, the mean, the covariance, and the correlation increase as k_{21} increases. The variance does not always increase uniformly, but instead often reaches a minimum at some migration level between 0.001 and 0.02. As a rule, however, the marginal distribution of X_2 always becomes more symmetric as k_{21} increases.

The impact of migration in this special model may be quantified as the difference in equilibrium population sizes. For the deterministic model, the difference in carrying capacities, $K_D = K_2 - K_1$, is a simple calculation. For example, for the standard migration rate of $k_{21} = 0.015$, one has

$K_D = 0.8902$. For the stochastic model in equilibrium, the difference

$$D = X_2 - X_1$$

is a random variable, whose distribution may be calculated directly with a simple transformation of the joint distribution $P(n_1, n_2)$. Figure 13.3 illustrates the distribution of D for $k_{21} = 0.015$. The first three cumulants of this distribution are $\kappa_{1D} = 0.9907$, $\kappa_{2D} = 4.919$ and $\kappa_{3D} = 0.375$, with index of skewness of only $\gamma_1 = 0.034$. It follows that a Normal approximation is very accurate for the distribution of D, and therefore one could construct prediction intervals for the size of D. For example, for this model with $a_1 = 0.30$ and $k_{21} = 0.015$, a 95% prediction interval for D, the difference in equilibrium population sizes is

$$0.991 \pm 4.347. \tag{13.6}$$

Figure 13.3. Equilibrium distribution for difference, D, between recipient and donor provinces in assumed NBDM model for AHB migration.

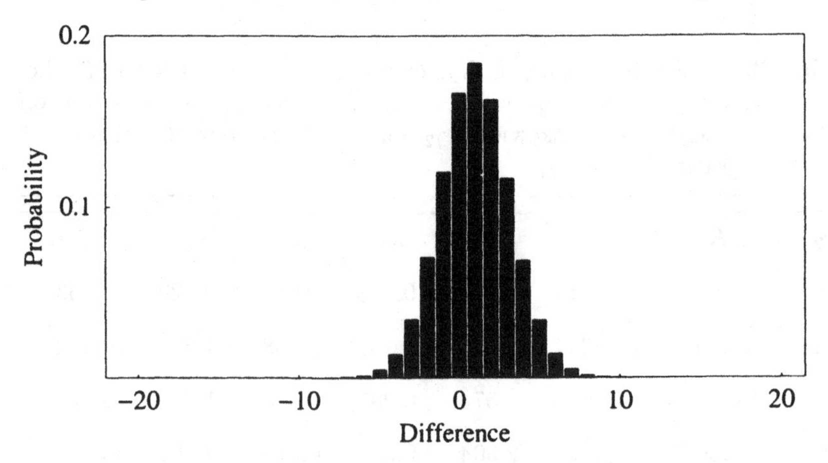

Five different families of models for $s = 1$ and $s = 2$, with intrinsic birthrates ranging from $a_1 = 0.05$ to 0.30, and with migration levels from $k_{21} = 0.001$ to 0.02 are investigated in [46]. It is shown that the means and variances of D may be estimated with a quadratic regression model in a_1 and k_{21}, in each case with R^2 exceeding 0.99. Moreover, in each case, the distribution of D is approximately Normal. It follows, therefore, that prediction intervals such as (13.6) are both accurate and easy to construct for a range of a_1 and k_{21} parameters for this migration model. We are hopeful that these simple methods would yield similar accuracy for many other similar models of biological spread, and such investigations are in progress.

13.3 Cumulant Functions

13.3.1 Derivation of Cumulant Approximations

Differential equations for approximating bivariate cumulant functions may be obtained relatively easily by applying previous methodology. Consider the modification discussed previously in Section 7.4, whereby the intensity functions for X_2 are modified to be polynomials for the whole range of X_2. With this modification, the intensity functions for the present model are:

$$\begin{aligned} f_{1,0} &= a_{11}X_1 - b_{11}X_1^{s+1} & f_{-1,0} &= a_{12}X_1 + b_{12}X_1^{s+1} \\ f_{0,1} &= a_{21}X_2 - b_{21}X_2^{s+1} & f_{0,-1} &= a_{22}X_2 + b_{22}X_2^{s+1} \\ f_{-1,1} &= k_{21}X_1 \end{aligned} \tag{13.7}$$

The partial differential equation for the mgf $M(\theta_1, \theta_2, t)$ obtained by substituting these into the operator equation in (9.34) is

$$\begin{aligned} \frac{\partial M}{\partial t} &= \sum_{i=1}^{2} \left[\left(e^{-\theta_i} - 1\right) a_{i1} + \left(e^{-\theta_i} - 1\right) a_{i2} \right] \frac{\partial M}{\partial \theta_i} \\ &+ \sum_{i=1}^{2} \left[\left(e^{\theta_i} - 1\right)(-b_{i1}) + \left(e^{-\theta_i} - 1\right) b_{i2} \right] \frac{\partial^{s+1} M}{\partial \theta_i^{s+1}} \\ &+ \left(e^{-\theta_1+\theta_2} - 1\right) k_{21} \frac{\partial M}{\partial \theta_1} \end{aligned} \tag{13.8}$$

The corresponding equation for the cgf is again obtained by the transformation $K = \log M$. For example, the result for $s = 1$ is:

$$\begin{aligned} \frac{\partial K}{\partial t} &= \sum_{i=1}^{2} \left[\left(e^{\theta_i} - 1\right) a_{i1} + \left(e^{-\theta_i} - 1\right) a_{i2} \right] \frac{\partial K}{\partial \theta_i} \\ &+ \sum_{i=1}^{2} \left[\left(e^{\theta_i} - 1\right)(-b_{i1}) + \left(e^{-\theta_i} - 1\right) b_{i2} \right] \left[\frac{\partial^2 K}{\partial \theta_i^2} + \left(\frac{\partial K}{\partial \theta_i} \right)^2 \right] \\ &+ \left(e^{-\theta_1+\theta_2} - 1\right) k_{21} \frac{\partial K}{\partial \theta_1} \end{aligned} \tag{13.9}$$

which is the generalization of the single density-dependent population result (6.16). Differential equations for the first and second order cumulants follow as

$$\begin{aligned} \dot{\kappa}_{10} &= \left(a_{1n} - b_{1n}\kappa_{10}\right)\kappa_{10} - b_{1n}\kappa_{20} \\ \dot{\kappa}_{01} &= k_{21}\kappa_{10} + \left(a_{2n} - b_{2n}\kappa_{01}\right)\kappa_{01} - b_{2n}\kappa_{02} \\ \dot{\kappa}_{20} &= \left(c_{1n} - d_{1n}\kappa_{10}\right)\kappa_{10} + \left(2a_{1n} - d_{1n} - 4b_{1n}\kappa_{10}\right)\kappa_{20} - 2b_{1n}\kappa_{30} \\ \dot{\kappa}_{02} &= \left(c_{2n} - d_{2n}\kappa_{01}\right)\kappa_{01} + \left(2a_{2n} - d_{2n} - 4b_{2n}\kappa_{01}\right)\kappa_{02} \\ &\quad + k_{21}\left(\kappa_{01} + 2\kappa_{11}\right) - 2b_{2n}\kappa_{03} \end{aligned} \tag{13.10}$$

$$\dot{\kappa}_{11} = (c_{1n} - 2b_{1n}\kappa_{10} + c_{2n} - 2b_{2n}\kappa_{01})\,\kappa_{11} + k_{21}\,(\kappa_{20} - \kappa_{10})$$
$$- b_{1n}\kappa_{21} - b_{2n}\kappa_{12},$$

with $a_{in} = a_{i1} - a_{i2} - k_{ji}$, $b_{in} = b_{i1} - b_{i2}$,

$$c_{in} = a_{i1} + a_{i2} + k_{ji} \quad \text{and} \quad d_{in} = b_{i1} - b_{i2} \tag{13.11}$$

An immediate generalization of the recommendation in Sections 6.5 and 7.5 for single populations is to solve equations (13.10) by obtaining first the additional four equations with third order cumulants from *Mathematica* [104]. One could then solve the truncated system of nine equations with all cumulants of fourth or higher order set to zero. The accuracy of this procedure on the marginal cumulants for $X_1(t)$ has been investigated in Chapter 6. The accuracy of the approximations for the transient cumulant functions involving $X_2(t)$ could be investigated as before by first solving the Kolmogorov equations. Of course, as a simpler, preliminary step, one could investigate the accuracy of the cumulants of the approximating equilibrium distribution, as in the previous section. Both of these types of investigations are illustrated in Chapter 14, but neither is pursued in this chapter. However an example follows to demonstrate a general application of this density-dependent growth model with migration.

13.3.2 Application to Muskrat Population Dynamics

We consider again the muskrat problem, and assume two populations with identical birth and death rate functions. Adjacent provinces are likely to have such identical rate functions as observed in Section 12.4. In particular we assume that an identical logistic model with $s = 1$ holds for both a donor (overijl) and a recipient (dr) province. The assumed common parameter estimates are the median values for the individual provinces in Table 6.2, specifically $a = 0.583$ and $b = 3.31 \times 10^{-5}$, as estimated for overijl. If these provinces also have identical starting values, then in the absence of migration they clearly would also have identical cumulant functions.

Three levels of migration are investigated in this model. One is $k_{21} = 0.10$, as suggested by the results for the linear migration model in Section 12.4. For another level, the rate is doubled to $k_{21} = 0.20$; for a third it is halved to $k_{21} = 0.05$. To adjust for migration from the first population, its overall intrinsic death rate, $a_2 = 0.25$, is partitioned into the migration rate, k_{21}, and a new intrinsic mortality rate, $a_{12} = a_2 - k_{21}$. The death rate for the second (recipient) province remains $a_{22} = 0.25$. The common parameters are $a_{11} = a_{21} = 0.833$, $b_{11} = b_{21} = 3.31 \times 10^{-5}$, and $b_{11} = b_{21} = 0$, with estimated initial values $X_1(0) = X_2(0) = 340$.

The approximate mean and variance functions for the first population (overijl) are given in Figure 13.4, which by construction are the same for all levels of migration. Figure 13.5 illustrates for the three levels of migration the comparative mean and variance functions for the second population

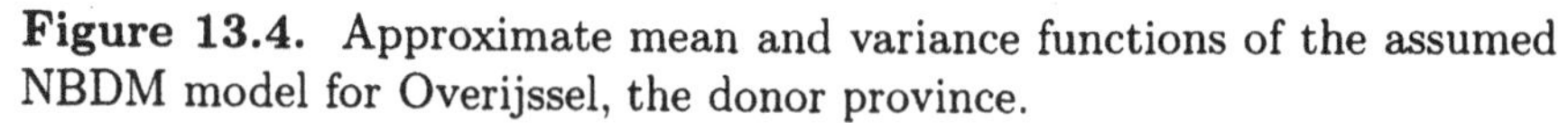

Figure 13.4. Approximate mean and variance functions of the assumed NBDM model for Overijssel, the donor province.

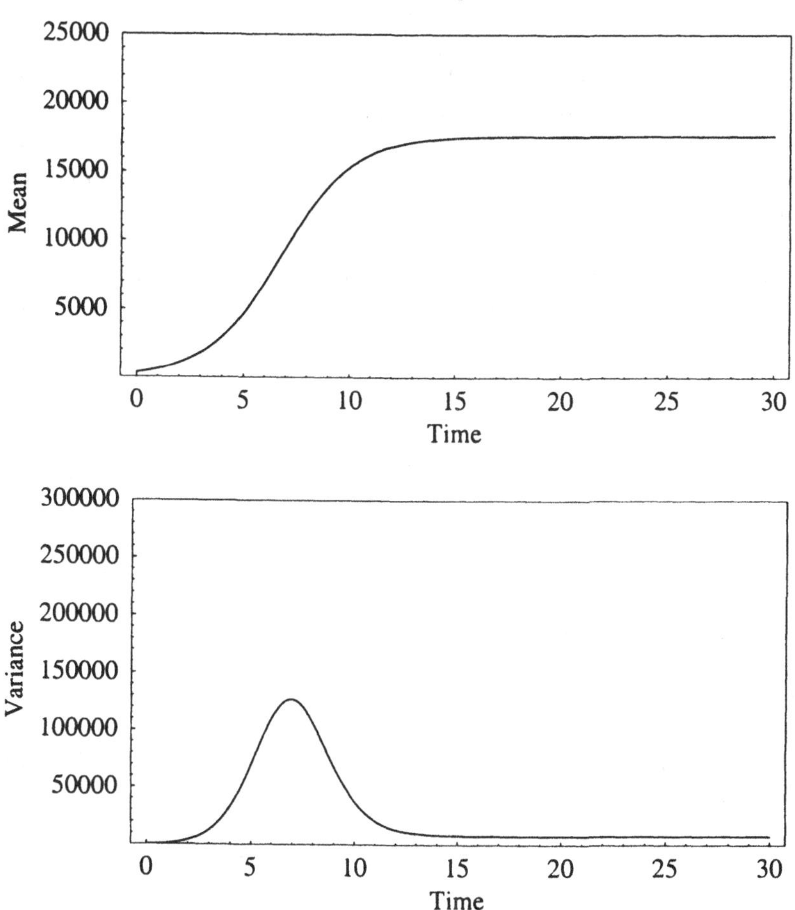

(dr) and also the covariance function. It is clear that the migration effect, even at the very low levels compared to the birth rate, has a substantial influence on the approximate cumulant functions of the recipient province.

Consider first the effect of migration on the mean value functions. The asymptotic mean value for overijl, the donor province with migration, is 17613; whereas the asymptotic means for the three levels of migration for dr, the recipient province, are 18317, 19133, and 20527 for $k = 0.05$, 0.10, and 0.20 respectively. The three proportional increases are 0.040, 0.086 and 0.165.

Though the variances are relatively small, there are striking effects of migration in the shape of the curves. For simplicity, we consider the standard deviation of population size. The peak standard deviation for the donor province in Figure 13.4 is 355.7; whereas the corresponding peaks for the three levels of migration in the recipient province in Figure 13.5 are 312.8, 294.7 and 286.7 respectively, for proportional decreases of 0.121, 0.171 and 0.194. The asymptotic standard deviation for overijl is 86.9, and the corresponding values for dr with the three migration levels are 83.7, 81.1, and 77.0, respectively, which represent decreases. It is interesting that the lower migration rates which give the smaller mean values also yield the larger standard deviations. Because the asymptotic variances are small, it follows that the differences between the deterministic carrying capacities and the corresponding asymptotic mean values of the equilibrium distributions are also small. Though these results apply only to the approximate cumulant functions, the approximate mean values are no doubt quite accurate and research is in progress to ascertain the accuracy of the second-order cumulant functions.

13.3.3 Conclusions

It is clear that the NBDM model developed in this chapter could be generalized along many different lines. The extensions to include immigration and two-way migration were previously discussed. The generalizations in the Appendix to Chapter 12, principly to multiple births and to nonexponential waiting times, could be readily incorporated using the approach in this monograph to obtain approximations for the underlying cumulant functions. The accuracy of such approximations would then require investigation.

Figure 13.5. Mean, variance and covariance functions of assumed NBDM model for Drente with four migration rates, k_{21}.

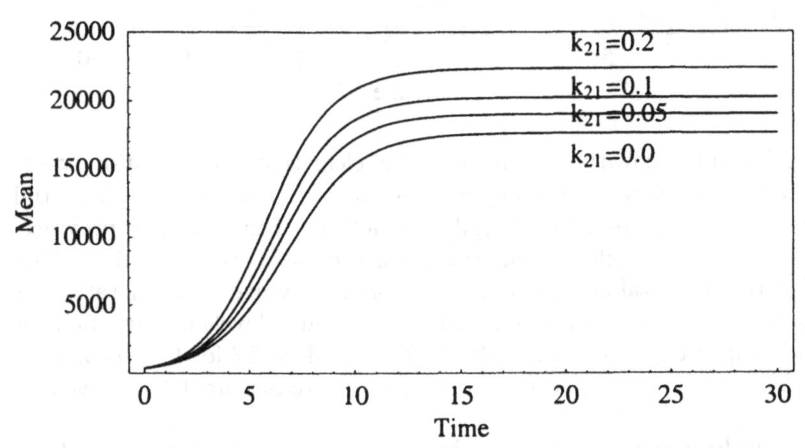

Figure 13.5. (Continued)

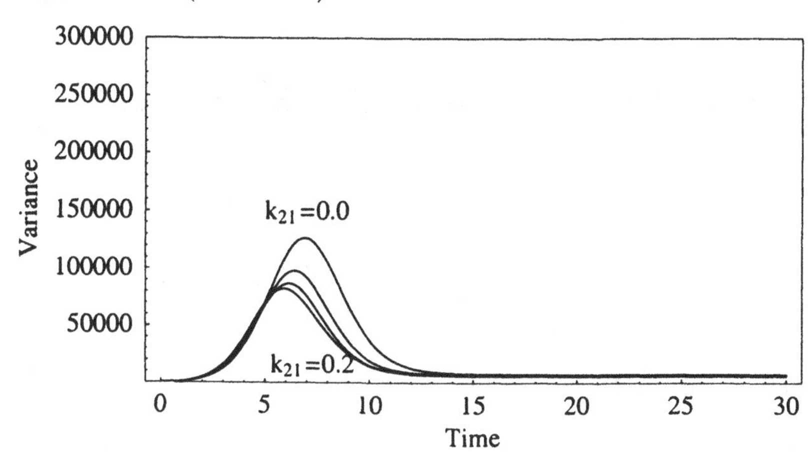

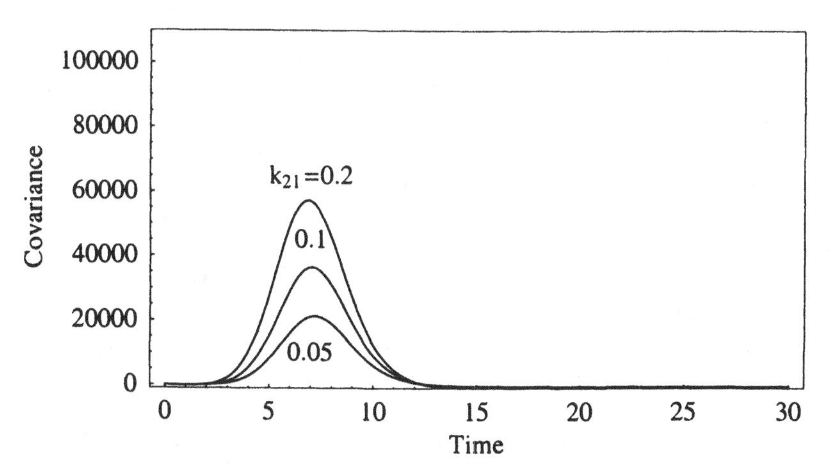

14

Nonlinear Host-Parasite Models

14.1 Introduction

Modeling the dynamics of interacting species is central in ecological theory, and there is a vast literature describing deterministic and stochastic models for such interactions (see e.g. [65, 66, 69, 73, 80]). A number of classic models for competition between multiple species and for host-parasite (and predator-prey) relationships have been formulated and widely investigated. This chapter investigates just one of these models, a modified Volterra host-parasite model, to illustrate the application of the methodology outlined in Chapter 9 which focuses on the underlying cumulant structure of the stochastic model.

We propose using host-parasite models for describing AHB/Varroa mite interactions. The reduced rate of northern AHB migration in the U.S. since 1990 [33] has motivated a number of studies reported in this momograph. Recent research suggests that parasitism by the Varroa mite (*Varroa jacobsoni*) is a major cause of the reduced migration rate. The mite was discovered invading managed European colonies (*Apis mellifera*) in 1992 in Texas, where it is estimated that the number of feral colonies has declined by more than 90% due to parasitization by Varroa mites [94]. This compares with a reported 75% loss of feral colonies in California [35], and a 71% loss in Arizona [36]. The devastating impact of the mites on managed European colonies since the 1970's is well known [38, 68]. Though recent research suggests that the AHB might be more resistant to the Varroa mite than the European bee [88], there is a consensus that it too is adversely affected by the mite, and that the observed slowing of AHB range expansion is due in part to the AHB/mite interaction.

We are not aware of any previous stochastic models for the dynamic interaction of these two particular species. Also, we are not aware of previous research for the applied *analysis* of the Volterra, or other related stochastic models, which is not constrained by the simplifying assumption of a bivariate Normal population size distribution. This is a crucial point in the present context, as our empirical observations and stochastic simulations of prototype bee/mite populations clearly suggest that the distributions of population size are skewed, as one might expect in ecological population process.

14.2 Proposed Host-Parasite Model

14.2.1 Model Definition

Consider now the populations of bee colonies, $X(t)$, and Varroa mites, $Y(t)$. The deterministic formulation of a so-called Volterra host-parasite model is reviewed in [80] and a stochastic analog of the model is proposed in [7]. Whittle [103] also considers this stochastic process, and suggests an analysis based on an assumed bivariate Normal distribution of population size. Research on the Volterra model is reviewed in [82], which illustrates this and related stochastic mdoels, and compares them to their deterministic analogs.

We propose for the present application extending the basic Volterra stochastic model in [7] to include the immigration of both host and parasite. The probabilities of unit changes in the populations in time interval t to $t + \Delta t$ are now assumed to be:

$$\begin{aligned}
&\text{Prob}\{\text{increase of 1 bee colony } |X(t), Y(t)\} \\
&\quad = \begin{cases} [I_1 + (a_1 - b_1 X)X]\,\Delta t & \text{for } X < a_1/b_1 \\ I_1 \Delta t & \text{otherwise} \end{cases} \\
&\text{Prob}\{\text{decrease of 1 bee colony } |X(t), Y(t)\} = (a_2 + c_1 Y)X\Delta t \\
&\text{Prob}\{\text{increase of 1 mite } |X(t), Y(t)\} = (I_2 + c_2 XY)\Delta t \qquad (14.1) \\
&\text{Prob}\{\text{decrease of 1 mite } |X(t), Y(t)\} = a_3 Y \Delta t.
\end{aligned}$$

These equations introduce two major qualitative changes from the single population nonlinear models in Chapter 7. For bee colonies the "per capita" death rate of $(a_2 + b_2 X)$ in (7.2) is replaced by $(a_2 + c_1 Y)$, which depends on the number of mites. Similarly, the per capita birth rate of mites is $c_2 X$, which is a simple linear function of the number of colonies.

Obviously many alternative models could be proposed, and more realism could be incorporated by introducing more parameters and richer structure. Some of these generalizations are discussed in Section 14.3. The modified Volterra model in (14.1), however, is sufficient for our purposes to capture the essence of the present bee/mite interaction with a minimum number of parameters.

The differential equations for the analogous deterministic model (for $X < a_1/b_1$) follow as

$$\begin{aligned}
\dot{X} &= I_1 + [(a_1 - b_1 X) - (a_2 + c_1 Y)]X \qquad (14.2) \\
\dot{Y} &= I_2 + (c_2 X - a_3)Y.
\end{aligned}$$

Their solution is illustrated subsequently.

The partial differential equation for $K(\theta_1, \theta_2, t)$, with modification as in (7.13), may be obtained using the operator equation in (9.34) with transformation (9.25) as:

$$
\begin{aligned}
\frac{\partial K}{\partial t} = & \left(e^{\theta_1} - 1\right)\left\{I_1 + a_1 \frac{\partial K}{\partial \theta_1} - b_1\left[\frac{\partial^2 K}{\partial \theta_1^2} + \left(\frac{\partial K}{\partial \theta_1}\right)^2\right]\right\} \\
& + \left(e^{-\theta_1} - 1\right)\left\{a_2 \frac{\partial K}{\partial \theta_1} + c_1\left[\frac{\partial^2 K}{\partial \theta_1 \partial \theta_2} + \left(\frac{\partial K}{\partial \theta_1}\right)\left(\frac{\partial K}{\partial \theta_2}\right)\right]\right\} \\
& + \left(e^{\theta_2} - 1\right)\left\{I_2 + c_2\left[\frac{\partial^2 K}{\partial \theta_1 \partial \theta_2} + \left(\frac{\partial K}{\partial \theta_1}\right)\left(\frac{\partial K}{\partial \theta_2}\right)\right]\right\} \\
& + \left(e^{-\theta_2} - 1\right) a_3 \frac{\partial K}{\partial \theta_2},
\end{aligned}
\tag{14.3}
$$

which generalizes (7.16). Substituting the power series expansion (9.26) into (14.3) yields the following differential equations for the first- and second-order cumulants:

$$
\begin{aligned}
\dot{\kappa}_{10}(t) &= I_1 + (a_1 - a_2 - b_1\kappa_{10} - c_1\kappa_{01})\kappa_{10} - b_1\kappa_{20} - c_1\kappa_{11} \\
\dot{\kappa}_{01}(t) &= I_2 - (a_3 - c_2\kappa_{10})\kappa_{01} + c_2\kappa_{11} \\
\dot{\kappa}_{20}(t) &= I_1 + (a_1 + a_2 - b_1\kappa_{10} + c_1\kappa_{01})\kappa_{10} + c_1(1 - 2\kappa_{10})\kappa_{11} \\
&\quad + (2a_1 - 2a_2 - b_1 - 4b_1\kappa_{10} - 2c_1\kappa_{01})\kappa_{20} - 2c_1\kappa_{21} \\
\dot{\kappa}_{02}(t) &= I_2 + (a_3 + c_2\kappa_{10})\kappa_{01} - 2(a_3 - c_2\kappa_{10})\kappa_{02} \\
&\quad + c_2(1 + 2\kappa_{01})\kappa_{11} + 2c_2\kappa_{12} \\
\dot{\kappa}_{11}(t) &= c_2\kappa_{20}\kappa_{01} - c_1\kappa_{10}\kappa_{02} + (a_1 - a_2 - a_3 - 2b_1\kappa_{10} \\
&\quad + c_2\kappa_{10} - c_1\kappa_{01})\kappa_{11} + (c_2 - b_1)\kappa_{21} - c_1\kappa_{12}.
\end{aligned}
\tag{14.4}
$$

As with the single population equations in Chapter 7, no exact solutions of the cumulant equations in (14.4) are possible because the equations for any j^{th} order joint cumulants always involve higher order cumulants. Hence, consider again cumulant truncation. Note that truncating the second-order cumulants in the first two equations yields the equivalent deterministic model in (14.2), which in this and most other applications differ considerably from the true mean value functions. Truncating the third-order cumulants in (14.2), as under the assumption of bivariate Normality, (see e.g. [27, 103]), and solving, gives accurate approximations for the means, but poor approximations for the second-order cumulants and, of course, imposes symmetry on the marginal distributions.

Suppose our objectives include solving for cumulant functions of up to third-order, which would include the skewness functions. Inasmuch as the model proposed in (14.1) is of power $s = 1$, we propose solving the system of equations including (14.4) and similar ones for the third- and fourth-order cumulants. The equations for the third-order cumulants are given in Appendix 14.4 and the equations for the fourth-order cumulants were also obtained using *Mathematica* [104]. Altogether, this procedure involves solving a system of 14 equations, with the fifth- and higher-order cumulants set to zero. An example follows.

14.2.2 Investigation of a Large-Scale Model.

Consider as an illustration a test case where the bees and mites reach a stochastic equilibrium. The consensus of experimenters is that such equilibrium is not possible for the European bee (*Apis mellifera*). However, the Indian bee (*Apis cerana*) is in equilibrium with the Varroa mite [68] and current research suggests that the Africanized bee subspecies (*Apis mellifera scutellata*) may also achieve such equilibrium.

For convenience, let the equilibrium value for bee colonies be $X^* = 20$ with immigration rate $I_1 = 0.20$/mo, which are similar to the parameters in the single population model in Section 7.2.2. The equilibrium value for mites is set at Y^*=2000, i.e. 100 mites/colony. Let the intrinsic rates be $a_1 = 0.30$, $a_2 = 0.02$, as before, and let $a_3 = 1$, which is consistent with empirical data.

The other parameters may be obtained by retaining $b_1 = 0.012$ from the single population model with $I_2 = 0$, and then solving for $c_1 = 2.5 \times 10^{-5}$ and $c_2 = 0.05$ from the deterministic equations by substituting $X^* = 20$ and $Y^* = 2000$ into (14.2) with $\dot{X}=\dot{Y}= 0$. In this model, most of the bee colony population control is attributable to the density-dependency factor $(b_1 X^2)$ rather than to the interaction with the mites. Hence we call this the 'lite-mite' model.

Our preferred model, however, is obtained by retaining the same equilibrium values, but balancing the nonlinear mite effect $(c_1 XY)$ in the bee death rate with the density dependency in the bee birthrate. Setting $b_1 = 0.007$ leads to the solutions $c_1 = 7.5 \times 10^{-5}$ and $c_2 = 0.05$, for which $b_1(X^*)^2$ in (14.1) is approximately equal to $c_1 X^* Y^*$. We call this the 'quite-a-mite' model because of the stronger bee/mite interaction effect. In order to illustrate the dynamics of mite population increase, the initial values for this model are chosen to be $X(0) = 20$ and $Y(0) = 20$, corresponding to an initial average of only one mite per hive.

Figure 14.1 gives the mean value functions for the two populations, as well as their corresponding deterministic solutions for this model. For the mite population, the solutions for the deterministic model and for the mean value of the stochastic model are quite different, both for the transient phase and in their equilibrium values. For the bee population, however, the equilibrium solutions for two curves are essentially equal. Both the deterministic and the mean value function solutions exhibit the characteristic damped oscillation pattern of classic host-parasite models.

Figure 14.1. Mean value functions and deterministic solutions for bee and mite populations of assumed quite-a-mite model.

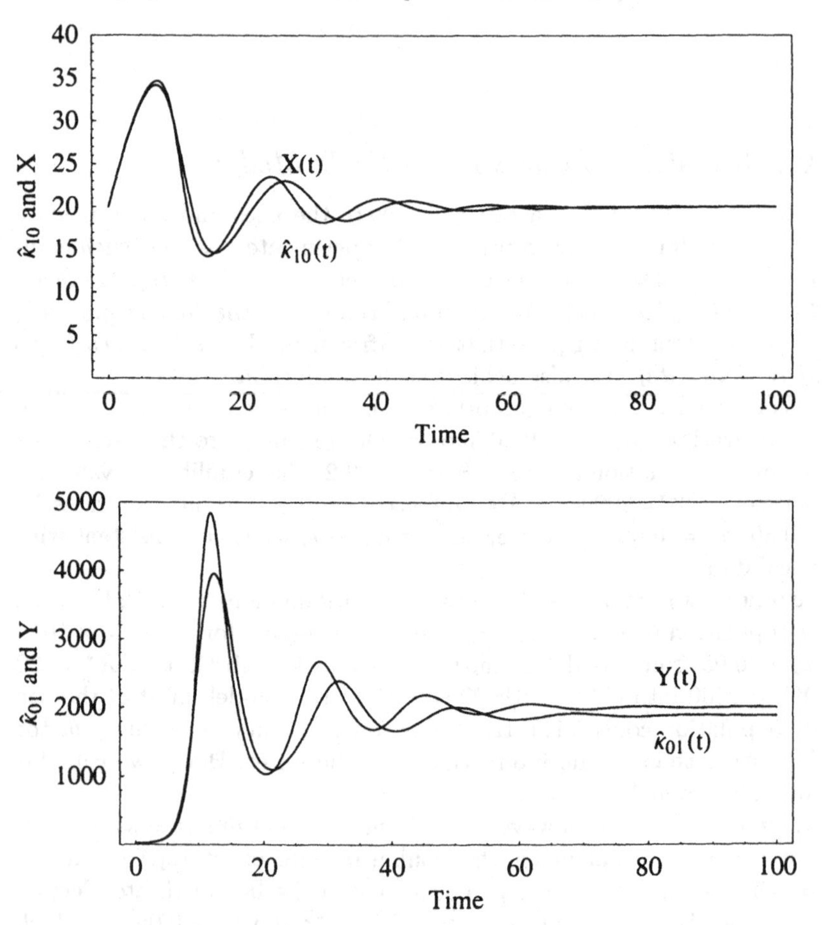

Figure 14.2. Variance and covariance functions for assumed quite-a-mite model.

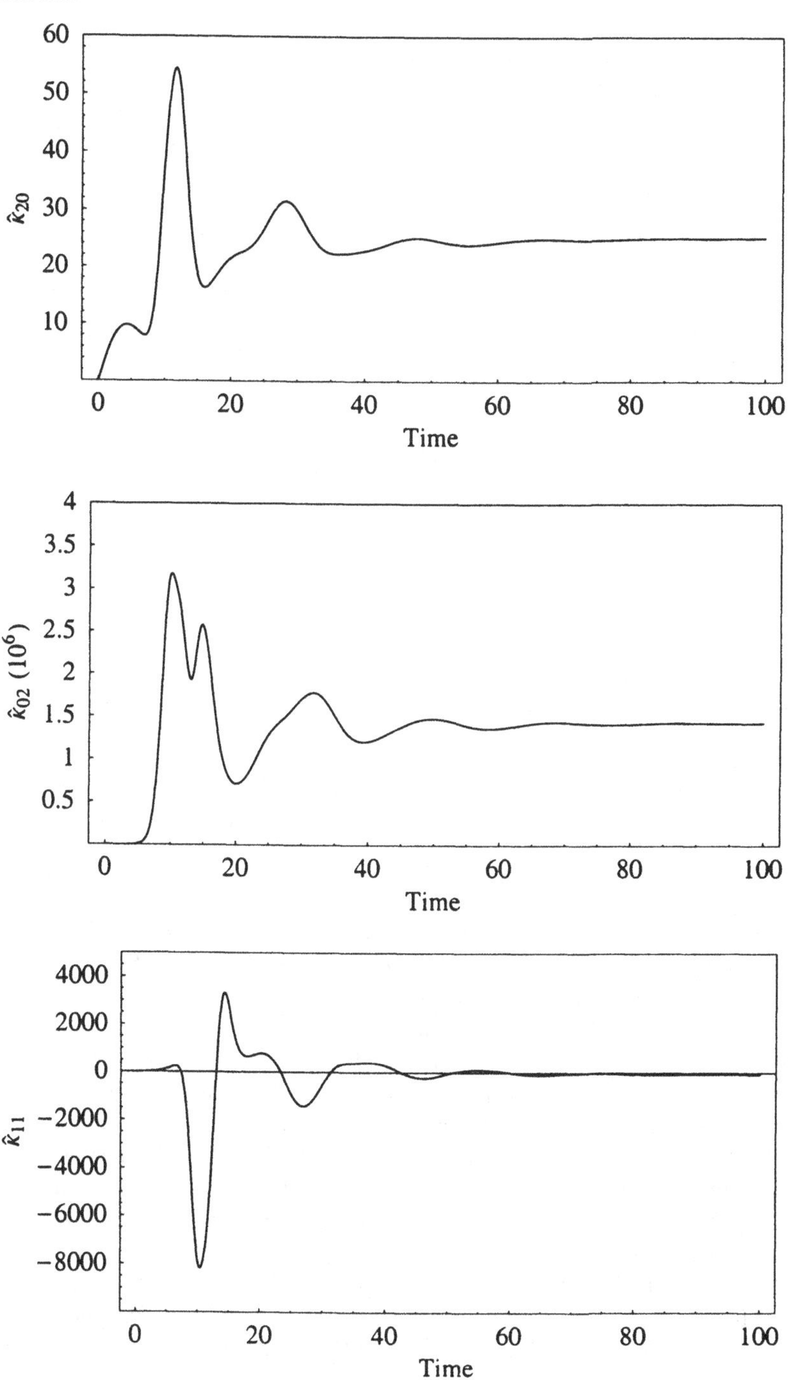

Figure 14.3. Skewness functions for assumed quite-a-mite model.

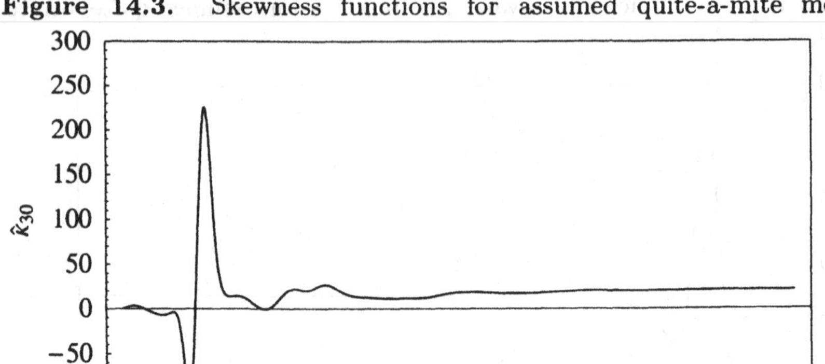

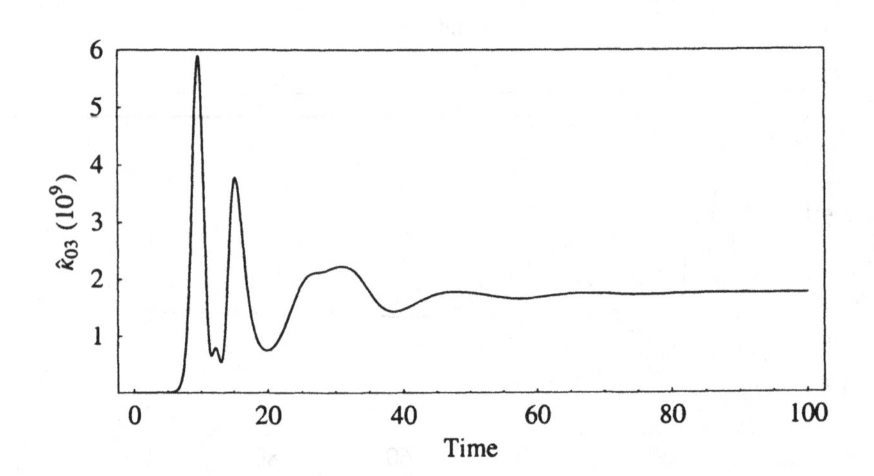

The second-order cumulant approximations, for the variances and covariance, are plotted in Figure 14.2 with the skewness results in Figure 14.3. The approximate equilibrium distributions have sizeable coefficients of variation, with $cv_1 = \kappa_{20}^{1/2}/\kappa_{10} = 0.25$ for bees and $cv_2 = \kappa_{02}^{1/2}/\kappa_{01} = 0.64$ for the mites. The approximate transient distributions are obviously very skewed at times during the initial period (for $t < 20$). The indices of skewness for the equilibrium distributions are $\gamma_1 = \kappa_{30}/(\kappa_{20})^{3/2} = 0.17$ which is small, and $\gamma_2 = \kappa_{03}/(\kappa_{02})^{3/2} = 1.03$, which implies visible skewness as one would expect for parasite populations of this type. An equilibrium distribution with skewness indices of roughly the same size as in this example is given in the next section.

Investigating the accuracy of these approximations is, however, problematic. In principle, one could solve a system of truncated Kolmogorov equations for the bivariate transient distributions. In this case though, with

X truncated at say 50 and Y at 4000, the set of equations for describing all possible joint probabilities is of size 200,000, which is intractable. One approach would be to use massive Monte Carlo simulation to determine the accuracy of these approximations. For the illustrative purposes of this paper, however, a smaller system is investigated.

14.2.3 Investigation of a Small-Scale Model.

To reduce the system size and thereby construct a more manageable model, let $X^* = 5$, consistent with USDA lab experiments, and $Y^* = 100$, i.e. an average of only 20 mites/colony. The intrinsic rates are retained at $a_1 = 0.30$, $a_2 = 0.02$, and $a_3 = 1$, with immigration rates set at $I_1 = 0.5$ and $I_2 = 10$. The solutions for the nonlinear coefficients, which yield rough parity for the $b_1(X^*)^2$ and $c_1X^*Y^*$ terms, are $b_1 = 0.038$, $c_1 = 0.0019$ and $c_2 = 0.18$. The model represents a small prototype case, for which the assumed virulence of the mites ensures that only 100 are needed to achieve equilibrium. Accordingly, it is called the 'dynamite' model. Though this low number of mites per colony is questionable biologically, the model is proposed to yield a convenient test case for investigation.

The deterministic solution of (14.2) for this model approaches its equilibrium value rapidly. Starting at $X(0) = Y(0) = 5$, the deterministic population sizes and a phase-plane plot of population size are given in Figure 14.4. Equilibrium, for all practical purposes, is achieved by $t = 25$ mo.

The analogous stochastic solution is strikingly different and obviously much richer. The Kolmogorov equations are solved by truncating $X_{\max} =$ 15 and $Y_{\max} = 450$. The two marginal equilibrium distributions of population size are given in Figure 14.5. One effect of truncating at $Y_{max} = 450$ is evident by the small accumulation of mass at $Y = 450$. The equilibrium distributions have means $\kappa_{10} = 4.648$ and $\kappa_{01} = 92.336$, variances $\kappa_{20} =$ 3.050 and $\kappa_{02} = 4133.1$ and skewnesses $\kappa_{30} = -0.29$ and $\kappa_{03} = 386.218$. The resulting substantial coefficients of variation, specifically $cv_1 = 0.38$ and $cv_2 = 0.70$, imply that the second-order terms are necessary for the accurate approximation of the mean value functions in (14.4). Stated in another way, the large cv's ensure that the mean value functions will differ measurably from their corresponding deterministic solutions. For example, note that even in the equilibrium case the deterministic solutions (of 5 and 100) differ by 7.0 and 7.7%, respectively, from the corresponding mean sizes obtained from the stochastic model. The indices of skewness are $\gamma_1 = 0.054$ and $\gamma_2 = 1.45$. The sizeable γ_2, which is visually apparent from the graph of the mite distribution in Figure 14.5, is an indication that the accurate approximation for the second-order cumulants from (14.4) requires the inclusion of the third-order cumulants.

Figure 14.4. Deterministic population sizes and phase-plane plot for assumed dynamite model.

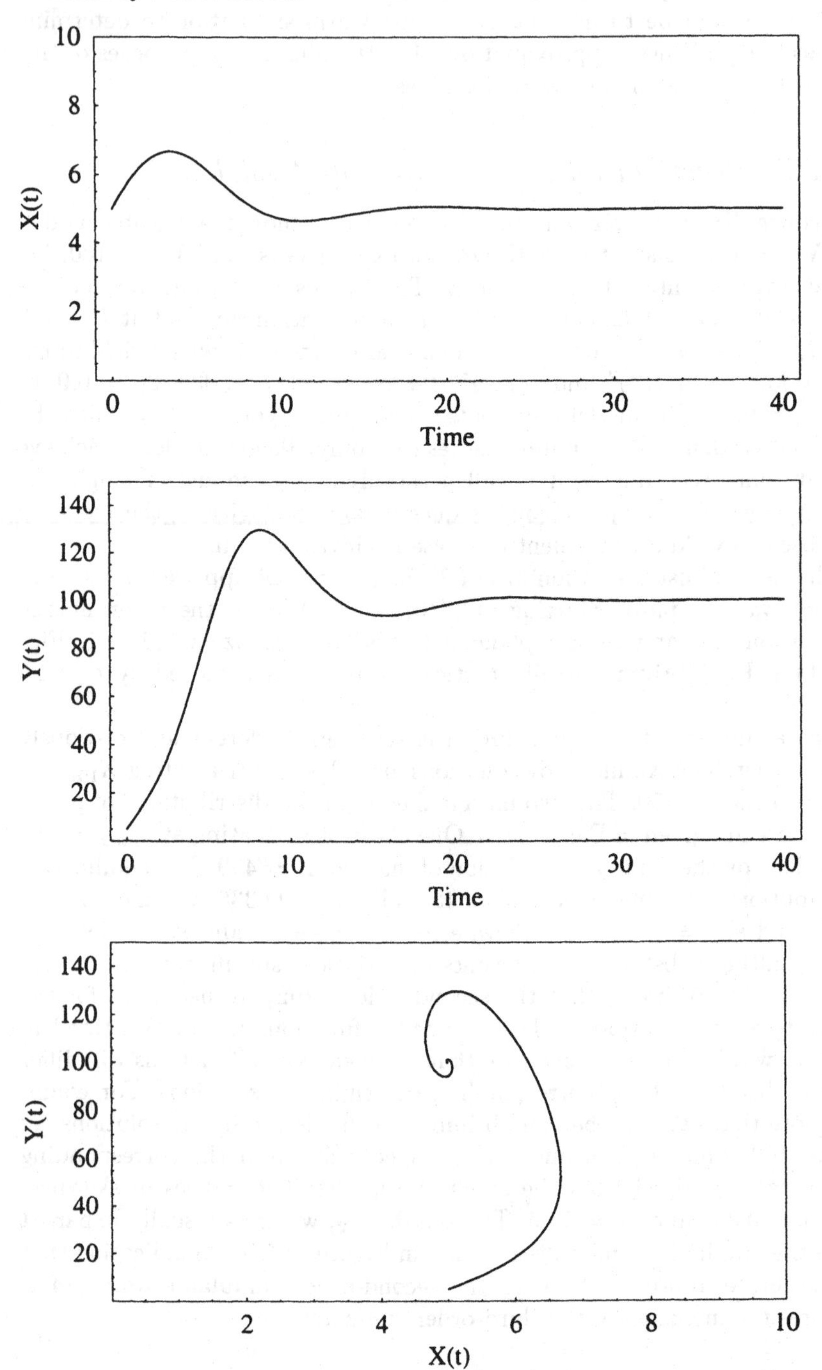

Figure 14.5. Marginal equilibrium distributions for bee and mite population sizes for assumed dynamite model.

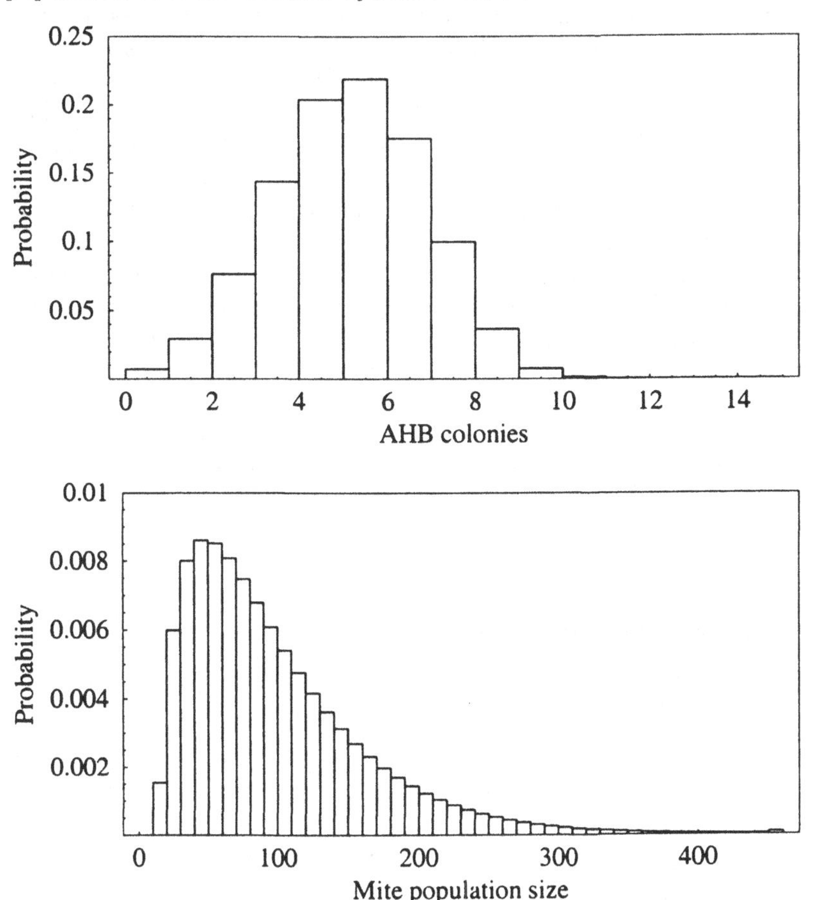

The cumulant function approximations were obtained as before, with the truncation of fifth- and higher-order cumulants. These solutions are compared in the subsequent figures with the "exact" cumulant functions obtained from the Kolmogorov solutions. In Figure 14.6 the approximate mean value functions are virtually indistinguishable visually from the exact solutions. As an example, the approximation errors for the two equilibrium distributions are only -0.5 and -0.1% respectively. The approximate second-order functions are reasonably accurate as evident in Figure 14.7. For example, the bivariate equilibrium distribution has exact cumulants $\kappa_{20} = 3.050$, $\kappa_{11} = 28.68$, and $\kappa_{02} = 4133.1$, and the approximation errors are -1.5, 6.3, and -2.8%, respectively. As apparent in Figure 14.8, some of the third-order functions have large errors. The joint cumulants for the equilibrium distribution are again representative. The exact values are $\kappa_{30} = -0.30$, $\kappa_{21} = -7.80$, $\kappa_{12} = 708.4$ and $\kappa_{03} = 386,218$, with approximation errors of -200, -2.8, -40, and -2.5% respectively. Though the 200% error might seem alarming, the absolute size of the skewness is tiny, hence the large percent error has little effect in the shape of the distribution.

As an illustration of the effect of these approximation errors in the cumulants, consider saddlepoint approximations of the equilibrium distribution for AHB colonies. For convenience, we use again the third-order approximation in (3.11). Table 14.1 compares the exact probabilities with their saddlepoint approximations based on the exact equilibrium cumulants. The approximation errors are moderate, with mean absolute size of 6.3% and poor accuracy only in the tails, for sizes 1 and 9. The exact probabilities are also compared in Table 14.1 to the saddlepoint approximation based on the approximate cumulants. The accuracy is about the same, in fact the mean absolute size is 5.2%, a bit smaller. The largest error is for size 0, in the left tail, which is a result of the large proportional increase in skewness $\widehat{\kappa}_{30}$ as compared to κ_{30}. This is only a single illustration using only the third-order saddlepoint approximation. Clearly further research is needed on the effect of approximation errors in cumulant functions on saddlepoint approximations.

Figure 14.6. Exact and approximate mean value functions for the bee and mite populations of the dynamite model.

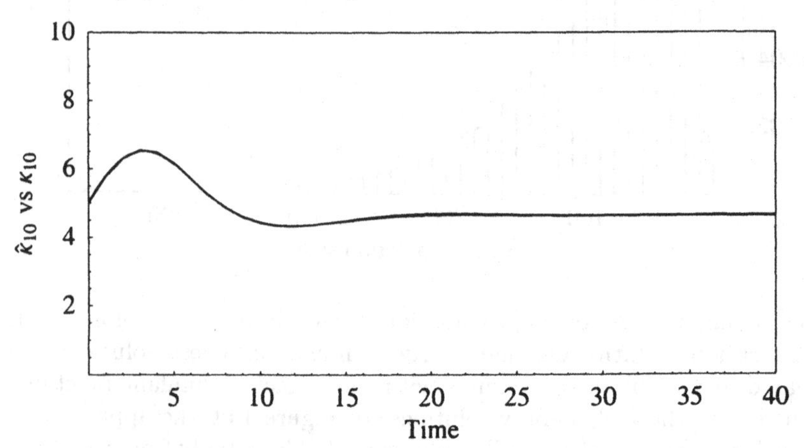

Figure 14.6. (Continued)

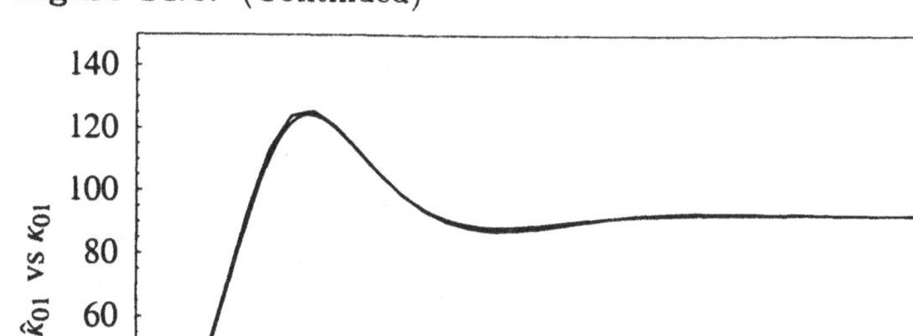

14.3 Conclusions and Future Research Directions

This chapter may be regarded as a progress report in our research applying stochastic models to describe the dynamic behavior of interacting African bee and Varroa mite populations. The paper establishes the feasibility of using cumulant truncation procedures to obtain approximate cumulant functions of stochastic bivariate processes, which may be very useful under conditions where the solution of the Kolmogorov equations is

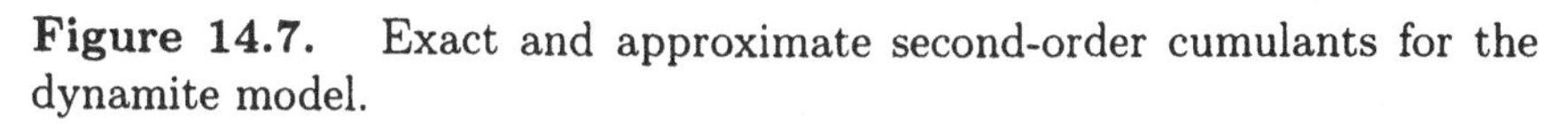

Figure 14.7. Exact and approximate second-order cumulants for the dynamite model.

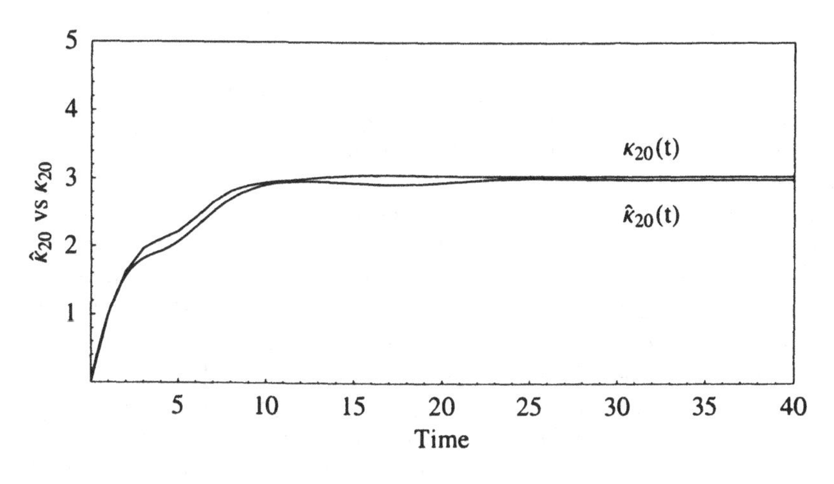

Figure 14.7. (Continued)

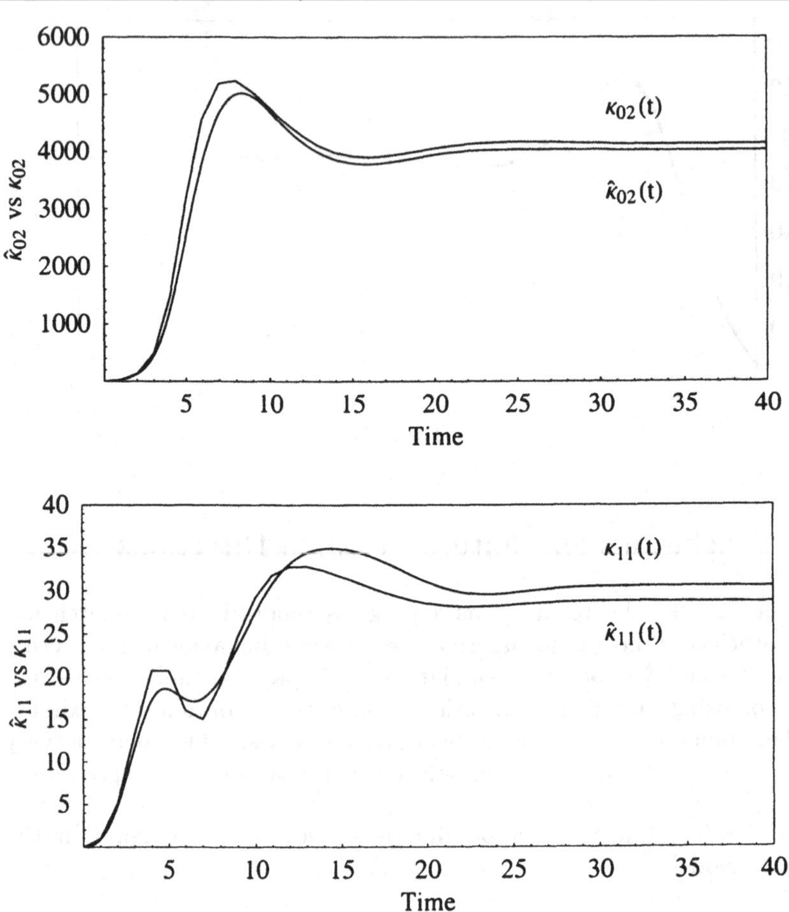

numerically intractable. Though the full analysis of our preferred, most realistic, model of bee and mite interactions is challenging due to the inherently large mite populations, a number of other applications in the literature involve smaller population sizes.

As an illustration, we examine in some detail a small prototype model for which the populations reach a stable stochastic equilibrium. In this example, the cumulant approximations for the first- and second-order cumulants are quite accurate. Research is in progress to investigate the accuracy of the larger, quite-a-mite, model using massive Monte Carlo simulation.

Research is also in progress developing the bivariate saddlepoint approximations for these models. Renshaw [84] extends a previous result in [83] by showing that the full saddlepoint approximation describes a bivariate Poisson distribution to within the accuracy of Stirling's approximation of $n!$. It also develops the full distribution approximation corresponding to the third-order cumulants in the AHB dispersal model [62]. The accuracy of using *truncated* sets of exact cumulants from bivariate stochastic processes is under current investigation.

Research is also in progress to extend the basic model defined in (14.1) in a number of ways. More realistic mite migration between colonies could be introduced using clustering [48] and colony compartmentalization. Multiple births, particularly of mites, and a mite age structure could be investigated, as illustrated in Appendix 12.5.1. The effects of environmental stochasticity could be studied, both as introduced through differential equations such as (14.4) directly [14], and through probability distributions for specified parameters [59]. Finally, spatial spread could be introduced to the model using a compartmental structure, as illustrated in Chapter 12. All of these extensions could be incorporated mathematically using generalizations of the partial differential equation in (14.3), from which cumulant equations similar to (14.4) follow. Hence the cumulant truncation methodology is also applicable to such extended host-parasite models for describing bee/mite interactions.

Figure 14.8. Exact and approximate third-order cumulants for the dynamite model.

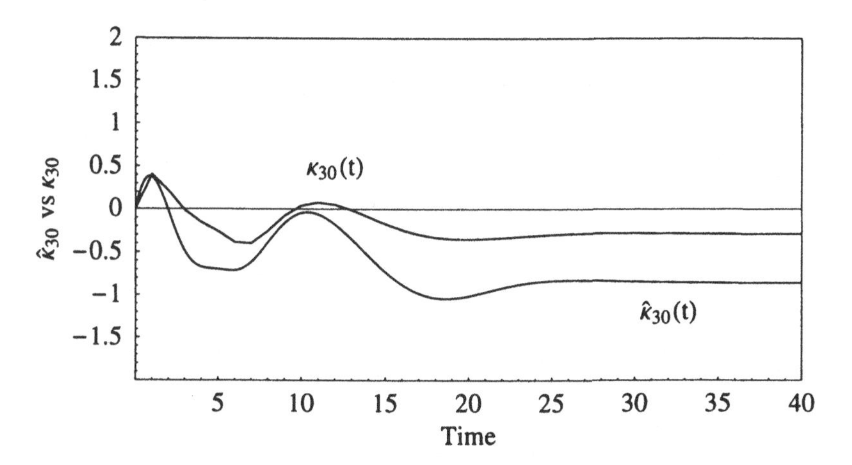

Figure 14.8 (Continued)

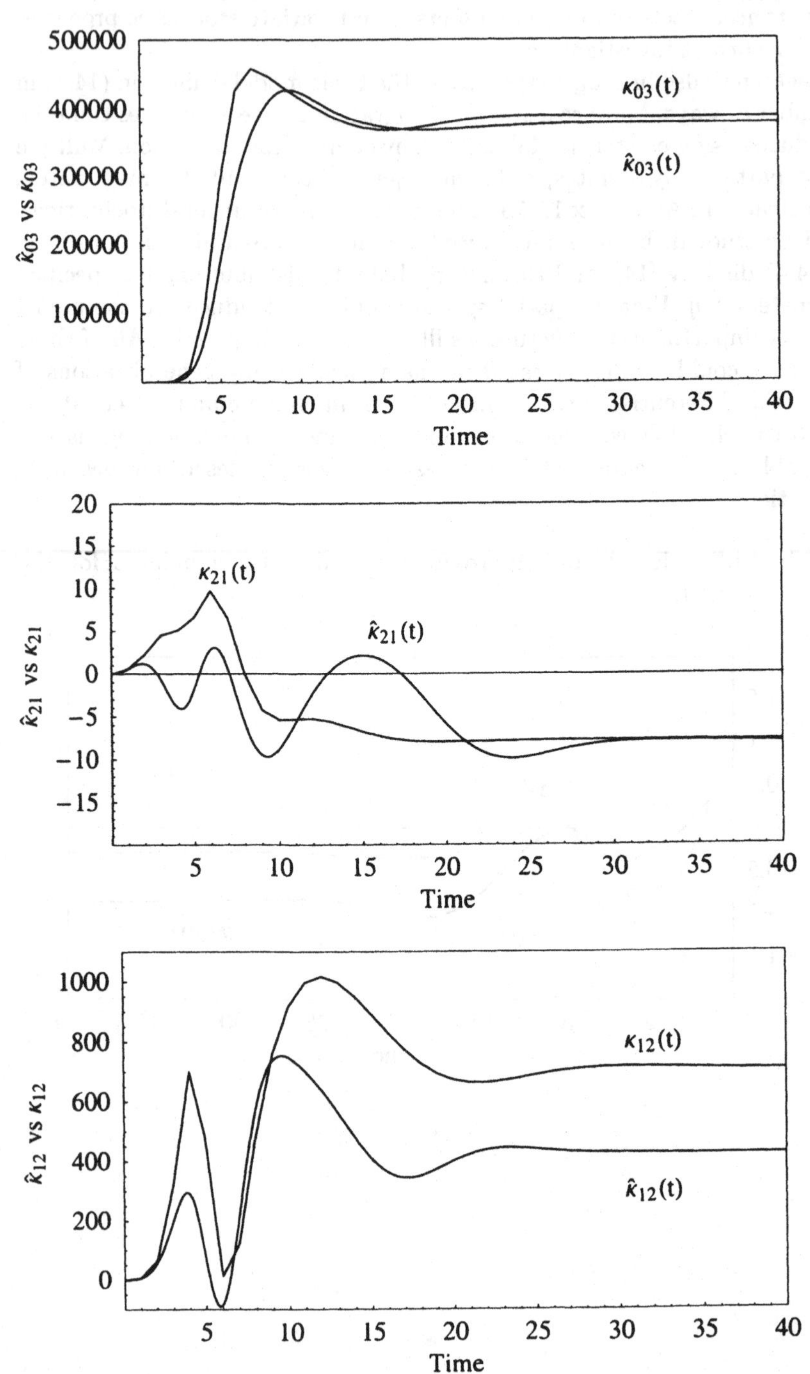

Table 14.1. Comparison of exact probabilities, $p_i(t)$, of AHB equilibrium size distribution with two saddlepoint approximations of order 3, one using exact cumulants, $\widehat{p}_i(t)$, and the other approximate cumulants from the truncation procedure, $\tilde{p}_i(t)$. Approximation errors in parentheses. * denotes no available value.

Size i	$p_i(t)$	$\widehat{p}_i(t)$	Error	$\tilde{p}_i(t)$	Error
0	.0071	.0073	(1.7%)	.0084	(17.3%)
1	.0294	.0264	(-10.1%)	.0278	(-5.5%)
2	.0764	.0718	(-6.0%)	.0717	(-6.1%)
3	.1436	.1439	(0.2%)	.1416	(-1.3%)
4	.2038	.2112	(3.6%)	.2101	(3.1%)
5	.2193	.2251	(3.6%)	.2290	(4.4%)
6	.1754	.1723	(-1.8%)	.1784	(1.7%)
7	.0996	.0937	(-5.9%)	.0958	(-3.8%)
8	.0366	.0358	(-2.1%)	.0338	(-7.7%)
9	.0074	.0095	(27.9%)	.0073	(-1.0%)
10	.0012	*		*	
11	.0002	*		*	

14.4 Appendix

14.4.1 Differential Equations for Third-Order Cumulants

The differential equations below complement those given in (14.4). The use of these in conjunction with (14.4) would enable one to approximate accurately the first- and second-order cumulants of the assumed host-parasite model.

$$\begin{aligned}\dot{\kappa}_{30}(t) &= I_1 + (a_1 - a_2 - b_1\kappa_{10} - c_1\kappa_{01})\kappa_{10} - c_1(1 - 3\kappa_{10})\kappa_{11} \\ &+ (3a_1 + 3a_2 - b_1 - 6b_1\kappa_{10} + 3c_1\kappa_{01} - 6c_1\kappa_{11} - 6b_1\kappa_{20})\kappa_{20} \\ &+ 3(a_1 - a_2 - b_1 - 2b_1\kappa_{10} - c_1\kappa_{01})\kappa_{30} + 3c_1(1 - \kappa_{10})\kappa_{21} \\ &- 3c_1\kappa_{31} - 3b_1\kappa_{40}\end{aligned}$$

$$\begin{aligned}\dot{\kappa}_{21}(t) &= (a_1 + a_2 - 2b_1\kappa_{10} + c_1\kappa_{01} + (2c_2 - 4b_1)\kappa_{20} - 2c_1\kappa_{11})\kappa_{11} \\ &+ c_1\kappa_{02}(\kappa_{10} - 2\kappa_{20}) + c_1(1 - 2\kappa_{10})\kappa_{12} + c_2\kappa_{01}\kappa_{30} \\ &+ (2a_1 - 2a_2 - a_3 - b_1 - (4b_1 - c_2)\kappa_{10} - 2c_1\kappa_{01})\kappa_{21} \\ &- 2c_1\kappa_{22} + (c_2 - 2b_1)\kappa_{31}\end{aligned}$$

$$\begin{aligned}\dot{\kappa}_{12}(t) &= (a_3 + c_2\kappa_{10} - c_1\kappa_{20} + 2(c_2 - b_1)\kappa_{11})\kappa_{11} \\ &+ c_2\kappa_{20}(\kappa_{01} + 2\kappa_{02}) + c_2(1 + 2\kappa_{01})\kappa_{21} \\ &+ (a_1 - a_2 - 2a_3 + 2(c_2 - b_1)\kappa_{10} - c_1\kappa_{01})\kappa_{12} - c_1\kappa_{10}\kappa_{03} \\ &- c_1\kappa_{13} + (2c_2 - b_1)\kappa_{22}\end{aligned}$$

$$\begin{aligned}\dot{\kappa}_{03}(t) &= I_2 - (a_3 - c_2\kappa_{10})\kappa_{01} + 3(a_3 + c_2\kappa_{10})\kappa_{02} - 3(a_3 - c_2\kappa_{10})\kappa_{03} \\ &+ c_2(1 + 3\kappa_{01} + 6\kappa_{02})\kappa_{11} + 3c_2(1 + \kappa_{01})\kappa_{12} \\ &+ 3c_2\kappa_{13}\end{aligned}$$

References

1 D. M. Allen and J. H. Matis, *KINETICA, a Program for Kinetic Modeling in Biological Sciences*, Department of Statistics, University of Kentucky, Lexington, KY, 1990.

2 D. M. Allen and J. H. Matis, Mean residence times and their standard errors for any interval of elapsed time, in J. Eisenfeld, D. S. Levine and M. Whitter (eds), *Biomedical Modeling and Simulation*, Elsevier Science Publ., New York, 1992.

3 D. H. Anderson, *Compartment Modeling and Tracer Kinetics*, Springer-Verlag, Berlin, 1983.

4 N. T. J. Bailey, *The Elements of Stochastic Processes*, Wiley, New York, 1964.

5 N. Balakrishnan, N. L. Johnson, and S. Kotz. A note on relationships between moments, central moments and cumulants from multivariate distributions, *Stat & Prob. Letters* 39: 49–54 (1998).

6 R. B. Banks, *Growth and Diffusion Phenomena*, Springer-Verlag, Berlin, 1994.

7 M. S. Bartlett, J. C. Gower, and P. H. Leslie, A comparison of theoretical and empirical results for some stochastic population models, *Biometrika* 47: 1–11 (1960).

8 D. M. Bates and D. G. Watts, *Nonlinear Regression Analysis & Its Applications*, Wiley, New York, 1988.

9 U. N. Bhat, *Elements of Applied Stochastic Processes*, Wiley, New York, 1972.

10 G. E. P. Box, *Robustness in Statistics*, Academic, New York, 1979.

11 G. E. P. Box, Some problems of statistics and everyday life. *J. Am. Statistical Assoc.* 74: 1–4 (1979).

12 C. L. Chiang, *An Introduction to Stochastic Processes and Their Applications*, R. E. Krieger, Melborne, FL, 1980.

13 Commission for Muskrat Eradication in the Netherlands, personal, communication of internal report.

14. B. Dennis, R. A. Desharnais, J. M. Cushing., and R. B. Constantino. Nonlinear demographic dynamics: Mathematical models, statistical methods, and biological experiments, *Ecol. Monographs.* 65:261–281 (1995).

15 W. C. Ellis, J. H. Matis, T. M. Hill and M. R. Murphy, Methodology for estimating digestion and passage kinetics of forages, in G. C. Fahey, Jr. (ed.), *Forage Quality, Evaluation and Utilization*, Am. Soc. of Agronomy, Madison, WI, 1994.

16 W. C. Ellis, D. P. Poppi, J. H. Matis, H. Lippke, T. M. Hill and F. M. Rouquette, Jr., Dietary-digestive-metabolic interactions determining the nutritive potential of ruminant diets, in H. J. G. Jung and G. C. Fahey, Jr. (eds.), *Nutritional Ecology of Herbivores*, Am. Soc. of Animal Sci, Savoy, IL, 1999.

17 C. S. Elton, *The Ecology of Invasions by Animals and Plants*, Methuen, London, 1958.

18 M. J. Faddy, Compartmental models with phase-type residence-time distributions, *Appl. Stoch. Models Data Anal.* 6:121–157 (1990).

19 M. Gibaldi and D. Perrier, *Pharmacokinetics* 2nd ed. Marcel Dekker, New York, 1982.

20 R. E. Goans, G. H. Weiss, S. A. Abrams, M. D. Perez, and A. L. Yergey. Calcium tracer kinetics show decreased irreversible flow to bone

in glucocorticoid treated patients, *Calcif. Tissue Int.* 56:533–535 (1995).

21 R. E. Goans, G. H. Weiss, N. E. Vieira, J. B. Sidbury, S. A. Abrams, and A. L. Yergey. Calcium kinetics in glycogen storage disease type 1a *Calcif. Tissue Int.* 59:449–453 (1996).

22 K. Godfrey, *Compartmental Models and Their Application*, Academic, London, 1983.

23 N. S. Goel and N. Richter-Dyn, *Stochastic Models in Biology*, Academic, New York, 1974.

24 W. E. Grant, J. H. Matis, and T. H. Miller, A stochastic compartmental model for migration of marine shrimp, *Ecol. Modelling* 54:1-15 (1991).

25 J. Hassan. Lesson plans. Science Academy of South Texas, Weslaco, TX. Personal communication.

26 R. Hengeveld, *Dynamics of Biological Invasions*, Chapman and Hall, London, 1989.

27 V. Isham. Assessing the variability of stochastic epidemics, *Math. Biosci.* 107: 209–224 (1991).

28 J. A. Jacquez, *Compartmental Analysis in Biology and Medicine*, 3rd ed., BioMedware, Ann Arbor, MI, 1996.

29 N. L. Johnson and S. Kotz, *Continuous Univariate Distributions*, Vol. 1, Wiley, New York, 1970.

30 N. L. Johnson, S. Kotz, and A. W. Kemp, *Univariate Discrete Distributions*, 2nd ed., Wiley, New York, 1992.

31 M. A. Johnson and M. R. Taaffe, Matching moments to phase distributions: Mixtures of Erlang distributions of common order, *Comm. Statist. Stochast. Models* 5:711–743 (1989).

32 M. Kac. Some mathematical models in science, *Science* 166:695-699 (1969).

33 J. K. Kaplan, Buzzing across the border, *Agricultural Research* 44:4–10 (1996).

34 D. G. Kendall, On the role of variable generation times in the development of a stochastic birth process, *Biometrika* 35:316-330 (1948).

35 B. Kraus and R. E. Page, Jr., Effect of *Varroa jacobsoni* (Mesostigmata: Varroidae) on feral *Apis mellifera* (Hymenoptera: Apidae) in California, *Environ. Entomol.* 24: 1473–1480 (1995).

36 G. M. Loper, A documented loss of feral bees due to mite infestations in S. Arizona, *Am. Bee J.* 135: 823–824 (1995).

37 G. Marion, E. Renshaw, and G. Gibson, Stochastic effects in a model of nematode infection in ruminants, *J. Math. Appl. Med. Biol.* 15: 97–116 (1998).

38 S. Martin, A population model for the ectoparasite mite *Varroa jacobsoni* in honey bee (*Apis melliferra*) colonies, *Ecological Modelling* 109:267–281 (1998).

39 J. H. Matis, Gamma time-dependency in Blaxter's compartmental model, *Biometrics* 28:597–602 (1972).

40 J. H. Matis, A paradox on compartmental models with Poisson immigration, *Am. Statistician* 27:84–85 (1973).

41 J. H. Matis, An introduction to stochastic compartmental models in pharmacokinetics, in A. Pecile and A. Rescigno (eds.), *Pharmacokinetics-Mathematical and Statistical Approaches to Metabolism and Distribution of Chemicals and Drugs*, Plenum Press, New York, 1988.

42 J. H. Matis, Statistical hammers and nails for predicting the spread of the Africanized honey (or "killer") bee, *Stats* 7:10–14 (1992).

43 J. H. Matis and H. O. Hartley, Stochastic compartmental analysis: Model and least squares estimation from time series data, *Biometrics* 27:77–102 (1971).

44 J. H. Matis and T. R. Kiffe, On approximating the moments of the equilibrium distribution of a stochastic logistic model, *Biometrics* 52:980–991 (1996).

45 J. H. Matis and T. R. Kiffe, Effects of immigration on some stochastic logistic models: A cumulant truncation analysis. *Theoretical Population Biol.* 56:139–161 (1999).

46 J. H. Matis and T .R. Kiffe, Migration effects in a stochastic multipopulation model for African bee population dynamics, *Environmental and Ecological Statistics* 4:301–319 (1997).

47 J. H. Matis and T. E. Wehrly, Modeling pharmacokinetic variability on the molecular level with stochastic compartmental systems, in M. Rowland, L. B. Sheiner and J. L. Steimer (eds), *Variability in Drug Therapy: Description, Estimation, and Control*, Raven, New York, 1985.

48 J. H. Matis and T. E. Wehrly, Compartmental models with multiple sources of stochastic variability: The one-compartment models with clustering, *Bull. Math. Biol.* 43: 651–644 (1981).

49 J. H. Matis and T. E. Wehrly, Generalized stochastic compartmental models with Erlang transit times, *J. Pharmacokin. Biopharm.* 18:589-607 (1991).

50 J. H. Matis and T. E. Wehrly, Compartmental models of ecological and environmental systems, in G. P. Patil and C. R. Rao (eds.), *Environmental Statistics*, Handbook of Statistics, vol. 12, Elsevier, New York, 1994.

51 J. H. Matis and T. E. Wehrly, A general approach to non-Markovian compartmental models, *J. Pharmacokin. Biopharm.* 26:437–456 (1998).

52 J. H. Matis, B. C. Patten, and G. C. White (eds), *Compartmental Analysis of Ecosystem Models*, Int. Coop. Publ. House, Fairland, MD (1979).

53 J. H. Matis, T. R. Kiffe, and R. Hengeveld, Estimating parameters for birth-death-migration models from spatio-temporal abundance data: Case of muskrat spread in the Netherlands. *J. Ag. Biol. Environ. Stat.* 1:40–59 (1996).

54 J. H. Matis, T. R. Kiffe, and G. W. Otis, Use of birth-death-migration processes for describing the spread of insect populations, *Environ. Ent.* 23:18–28 (1994).

55 J. H. Matis, T. R. Kiffe, and P. R. Parthasarathy, Using density-dependent birth-death-migration models for analyzing muskrat spread in the Netherlands. *Jour. Indian Soc. Ag. Statistics* 49:139–156 (1997).

56 J. H. Matis, T. R. Kiffe, and P. R. Parthasarathy, On the cumulants of population size for the stochastic power law logistic model. *Theor. Popul. Biol.* 53:16–29 (1998).

57 J. H. Matis, T. H. Miller, and D. M. Allen, Stochastic models of bioaccumulation, in M. C. Newman and A. W. McIntosh (eds), *Metal Ecotoxicology Concepts and Applications*, Lewis, Chelsea, MI, 1991.

58 J. H. Matis, W. L. Rubink and M. Makela, Use of the gamma distribution for predicting arrival times of invading insect populations, *Environ. Entomol.* 21:436–440 (1992).

59 J. H. Matis, T. E. Wehrly, and W. C. Ellis, Some generalized stochastic compartment models for digesta flow, *Biometrics* 45:703–720 (1989).

60 J. H. Matis, T. E. Wehrly, and C. M. Metzler, On some stochastic formulations and related statistical moments in pharmacokinetic models, *J. Pharmacokin. Biopharm.* 11:77–92 (1983).

61 J. H. Matis, W. E. Grant, and T. H. Miller, A semi-Markov process model for migration of marine shrimp, *Ecol. Model.* 60:167–184 (1992).

62 J. H. Matis, Q. Zheng, and T. R. Kiffe, Describing the spread of biological populations using stochastic compartmental models with births, *Math. Biosciences* 126:215–247 (1995).

63 J. H. Matis, Q. Zheng, and W. L. Rubink, Modeling spread of the Africanized honey bee–a case study, in *Proc. Section on Stat. and Environ.* Am. Stat. Assoc., Alexandria, VA, 1991.

64 J. H. Matis, T. E. Wehrly, D. M. Allen, and G. W. Otis, On using stochastic compartmental models for describing insect dispersal: 1. The case of univariate distributions from Markov process model, in *Biomedical Modeling and Simulation*, J. Eisenfeld, D. S. Levine and M. Whitten (eds.), Elsevier, New York, 1992.

65 R. M. May, *Stability and Complexity in Model Ecosystems*, University Press, Princeton, NJ, 1973.

66 J. Maynard Smith, *Models in Ecology.* University Press, Cambridge, UK, 1974.

67 Micromath Inc., *Scientist, for Experimental Data Fitting*, Micromath, Salt Lake City, UT, 1995.

68 B. Mobus and C. de Bruyn, *The New Varroa Handbook.* Northern Bee Books, Mytholmroyd, UK, 1993.

69 J. D. Murray, *Mathematical Biology* 2nd ed., Springer-Verlag, New York, 1993.

70 J. Neter, M. H. Kutner, C. J. Nachtsheim, and W. Wasserman, *Applied Linear Regression Models*, 3rd ed., Irwin, Chicago, IL, 1996.

71 M. F. Neuts, *Matrix-Geometric Solutions in Stochastic Models: An Algorithmic Approach*, Johns Hopkins Univ. Press, Baltimore, MD, 1981.

72 M. C. Newman and D. K. Doubet, Size-dependence of mercury accumulation kinetics in the mosquito fish, *Gambusia affinis*, *Arch. Environ. Contam. Toxicol.* 18:819–825 (1989).

73 R. M. Nisbet and W. S. C. Gurney, *Modelling Fluctuating Populations*, Wiley, New York, 1982.

74 J. K. Ord, G. P. Patil, and C. Taillie (eds). *Statistical Distributions in Ecological Work.* Int. Coop. Publ. House, Fairland, MD, 1979.

75 G.W. Otis, *The Swarming Biology and Population Dynamics of the Africanized Honey Bee.* Ph.D. Diss, Univ. Kansas, Manhattan, KS, 1980.

76 G.W. Otis, Population biology of the Africanized honey bee, in *The "African" Honey Bee,* M. Spivak, D.J.C. Fletcher and M.D. Breed, (eds.), Westview, Boulder, CO, 1991.

77 P. R. Parthasarathy, R. B. Lenin, and J. H. Matis, On the numerical solution of transient probabilities of the stochastic power law logistic model, *Nonlinear Analysis* 37:677-688 (1999).

78 E. Parzen, *Stochastic Processes*, Holden-Day, San Francisco, 1962.

79 B. C. Patten and J. H. Matis, The water environs of Okefenokee Swamp: An application of static linear environ analysis, *Ecol. Modelling* 16:1–50 (1982).

80 E. C. Pielou, *Mathematical Ecology,* Wiley, New York, 1977.

81 E. Renshaw, A survey of stepping-stone models in population dynamics, *Adv. Appl. Prob.* 18:581-627 (1986).

82 E. Renshaw, *Modeling Biological Populations in Space and Time*, Cambridge Univ. Press, New York, 1991.

83 E. Renshaw, Saddlepoint approximations for stochastic processes with truncated cumulant generating functions. *J. Math. Appl. Med. Biol.* 15: 1–12 (1998).

84 E. Renshaw, Applying the saddlepoint approximation to bivariate stochastic processes. Manuscript (2000).

85 A. Rescigno and G. Segre, *Drug and Tracer Kinetics*, Blaisdell, Waltham, MA, 1996.

86 T. E. Rinderer and R. L. Hellmich II, The processes of Africanization, in M. Spivak, D. J. C. Fletcher, and M. D. Breed (eds), *The "African" Honey Bee*, Westview, Boulder, CO, 1991.

87 G. H. Rowell, M. E. Makela, J. D. Villa, J. H. Matis, J. M. Labougle and R. R. Taylor, Jr., Invasive dynamics of Africanized honeybees in North America, *Naturwissenschaften* 79:281–283 (1992).

88 W. L. Rubink, P. Luevano-Martinez, E. A. Sugden, W. T. Wilson, and A. M. Collins, Subtropical *Apis mellifera* swarming dynamics and Africanization rates in Northeastern Mexico and Southern Texas, *Ann. Entomol. Soc. Am.* 87:243–251 (1996).

89 J. G. Skellam, Random dispersal in theoretical populations, *Biometrika* 38:196-218 (1951). teme90 P. J. Smith, A recursive formulation of the old problem of obtaining moments from cumulants and vice versa, *Am. Statistician* 49:217–218 (1995).

91 A. Spacie and J. L. Hamelink, Bioaccumulation, in G. M. Rand and S. R. Petrocelli (eds.), *Fundamentals of Aquatic Toxicology*, Hemisphere Publ., Washington, DC, 1985.

92 M. Spivak, D. J. C. Fletcher, and M. D. Breed. *The "African" Honey Bee*, Westview, Boulder, CO, 1991.

93 A. Stuart and J. K. Ord, *Kendall's Advanced Theory of Statistics.* Vol. 1. Distribution Theory, 5th ed. Oxford Press, New York, 1987.

94 Texas Agricultural Experiment Station, *The Texas Honey Bee Research and Management Plan.* College Station, TX, 1998.

95 J. Thomas, Texas Agricultural Experiment Station, personal communication. Also given in [42].

96 F. van den Bosch, J. A. J. Metz, and O. Dickmann, The velocity of spatial population expansion, *J. Math. Biol.* 28:529-565 (1990).

97 F. van den Bosch, R. Hengeveld, and J. A. J. Metz, Analyzing the velocity of animal range expansion, *J. Biogeogr.* 19:135-150 (1992).

98 W. Y. Velez, Integration of research and education, *Notices of Am. Math. Soc* 43:1142–1146 (1996).

99 T. E. Wehrly, J. H. Matis, and G. W. Otis, Approximating multivariate distributions in stochastic models of insect population dynamics, in G. P. Patil, C. R. Rao and N. P. Ross, (eds.), *Multivariate Environmental Statistics*, Elsevier, New York, 1994.

100 G. H. Weiss, R. E. Goans, M. Gitterman, S. A. Abrams, N. E. Vieira and A. L. Yergey. A non-Markovian model for calcium kinetics, *J. Pharmacokin. Biopharm.* 22:367–379 (1994).

101 J. K. Westbrook, W. W. Wolf, P. D. Lingren, J. R. Raulston, J. D. Lopez, J. H. Matis, R. S. Eyster, J. F. Esquivel, and P. G. Schlieder, Early-season migratory flights of corn earworm. *Environ. Entomol.* 26:12–20 (1997).

102 M. E. Wise, The need for rethinking on both compartments and modelling, in J. H.Matis, B. C. Patten and G. C. White (eds), *Compartmental Analysis of Ecosystem Models*, Int. Coop. Publ. House, Fairland, MD 1979.

103 P. Whittle, On the use of the Normal approximation in the treatment of stochastic processes. *J. Roy. Statist. Soc.* B. 19:268–281 (1957).

104 S. Wolfram, *The Mathematica Book*, 4th Ed. Wolfram Media/Cambridge Univ. Press. Champaign, IL, 1999.

105 L. J. Young and J. H. Young, *Statistical Ecology*, Kluwer, Boston, MA, 1998.

106 J. Yu, A non-Markovian compartment model approach with calcium clearance data. Manuscript. Personal communication. (2000).

107 Q. Zheng, Computing relations between statistical moments and cumulants, Manuscript. Personal communication (2000).

108 Q. Zheng and J. H. Matis, Some applications, properties and conjectures for higher-order cumulants of a Markovian stepping-stone model, *Comm. Stat - Theory Meth.* 22:3305-3319 (1993).

Index

Lecture Notes in Statistics

For information about Volumes 1 to 71, please contact Springer-Verlag

Vol. 72: J.K. Lindsey, The Analysis of Stochastic Processes using GLIM. vi, 294 pages, 1992.

Vol. 73: B.C. Arnold, E. Castillo, J.-M. Sarabia, Conditionally Specified Distributions. xiii, 151 pages, 1992.

Vol. 74: P. Barone, A. Frigessi, M. Piccioni, Stochastic Models, Statistical Methods, and Algorithms in Image Analysis. vi, 258 pages, 1992.

Vol. 75: P.K. Goel, N.S. Iyengar (Eds.), Bayesian Analysis in Statistics and Econometrics. xi, 410 pages, 1992.

Vol. 76: L. Bondesson, Generalized Gamma Convolutions and Related Classes of Distributions and Densities. viii, 173 pages, 1992.

Vol. 77: E. Mammen, When Does Bootstrap Work? Asymptotic Results and Simulations. vi, 196 pages, 1992.

Vol. 78: L. Fahrmeir, B. Francis, R. Gilchrist, G. Tutz (Eds.), Advances in GLIM and Statistical Modelling: Proceedings of the GLIM92 Conference and the 7th International Workshop on Statistical Modelling, Munich, 13-17 July 1992. ix, 225 pages, 1992.

Vol. 79: N. Schmitz, Optimal Sequentially Planned Decision Procedures. xii, 209 pages, 1992.

Vol. 80: M. Fligner, J. Verducci (Eds.), Probability Models and Statistical Analyses for Ranking Data. xxii, 306 pages, 1992.

Vol. 81: P. Spirtes, C. Glymour, R. Scheines, Causation, Prediction, and Search. xxiii, 526 pages, 1993.

Vol. 82: A. Korostelev and A. Tsybakov, Minimax Theory of Image Reconstruction. xii, 268 pages, 1993.

Vol. 83: C. Gatsonis, J. Hodges, R. Kass, N. Singpurwalla (Editors), Case Studies in Bayesian Statistics. xii, 437 pages, 1993.

Vol. 84: S. Yamada, Pivotal Measures in Statistical Experiments and Sufficiency. vii, 129 pages, 1994.

Vol. 85: P. Doukhan, Mixing: Properties and Examples. xi, 142 pages, 1994.

Vol. 86: W. Vach, Logistic Regression with Missing Values in the Covariates. xi, 139 pages, 1994.

Vol. 87: J. Müller, Lectures on Random Voronoi Tessellations.vii, 134 pages, 1994.

Vol. 88: J. E. Kolassa, Series Approximation Methods in Statistics. Second Edition, ix, 183 pages, 1997.

Vol. 89: P. Cheeseman, R.W. Oldford (Editors), Selecting Models From Data: AI and Statistics IV. xii, 487 pages, 1994.

Vol. 90: A. Csenki, Dependability for Systems with a Partitioned State Space: Markov and Semi-Markov Theory and Computational Implementation. x, 241 pages, 1994.

Vol. 91: J.D. Malley, Statistical Applications of Jordan Algebras. viii, 101 pages, 1994.

Vol. 92: M. Eerola, Probabilistic Causality in Longitudinal Studies. vii, 133 pages, 1994.

Vol. 93: Bernard Van Cutsem (Editor), Classification and Dissimilarity Analysis. xiv, 238 pages, 1994.

Vol. 94: Jane F. Gentleman and G.A. Whitmore (Editors), Case Studies in Data Analysis. viii, 262 pages, 1994.

Vol. 95: Shelemyahu Zacks, Stochastic Visibility in Random Fields. x, 175 pages, 1994.

Vol. 96: Ibrahim Rahimov, Random Sums and Branching Stochastic Processes. viii, 195 pages, 1995.

Vol. 97: R. Szekli, Stochastic Ordering and Dependence in Applied Probability. viii, 194 pages, 1995.

Vol. 98: Philippe Barbe and Patrice Bertail, The Weighted Bootstrap. viii, 230 pages, 1995.

Vol. 99: C.C. Heyde (Editor), Branching Processes: Proceedings of the First World Congress. viii, 185 pages, 1995.

Vol. 100: Wlodzimierz Bryc, The Normal Distribution: Characterizations with Applications. viii, 139 pages, 1995.

Vol. 101: H.H. Andersen, M.Højbjerre, D. Sørensen, P.S.Eriksen, Linear and Graphical Models: for the Multivariate Complex Normal Distribution. x, 184 pages, 1995.

Vol. 102: A.M. Mathai, Serge B. Provost, Takesi Hayakawa, Bilinear Forms and Zonal Polynomials. x, 378 pages, 1995.

Vol. 103: Anestis Antoniadis and Georges Oppenheim (Editors), Wavelets and Statistics. vi, 411 pages, 1995.

Vol. 104: Gilg U.H. Seeber, Brian J. Francis, Reinhold Hatzinger, Gabriele Steckel-Berger (Editors), Statistical Modelling: 10th International Workshop, Innsbruck, July 10-14th 1995. x, 327 pages, 1995.

Vol. 105: Constantine Gatsonis, James S. Hodges, Robert E. Kass, Nozer D. Singpurwalla(Editors), Case Studies in Bayesian Statistics, Volume II. x, 354 pages, 1995.

Vol. 106: Harald Niederreiter, Peter Jau-Shyong Shiue (Editors), Monte Carlo and Quasi-Monte Carlo Methods in Scientific Computing. xiv, 372 pages, 1995.

Vol. 107: Masafumi Akahira, Kei Takeuchi, Non-Regular Statistical Estimation. vii, 183 pages, 1995.

Vol. 108: Wesley L. Schaible (Editor), Indirect Estimators in U.S. Federal Programs. viii, 195 pages, 1995.

Vol. 109: Helmut Rieder (Editor), Robust Statistics, Data Analysis, and Computer Intensive Methods. xiv, 427 pages, 1996.

Vol. 110: D. Bosq, Nonparametric Statistics for Stochastic Processes. xii, 169 pages, 1996.

Vol. 111: Leon Willenborg, Ton de Waal, Statistical Disclosure Control in Practice. xiv, 152 pages, 1996.

Vol. 112: Doug Fischer, Hans-J. Lenz (Editors), Learning from Data. xii, 450 pages, 1996.

Vol. 113: Rainer Schwabe, Optimum Designs for Multi-Factor Models. viii, 124 pages, 1996.

Vol. 114: C.C. Heyde, Yu. V. Prohorov, R. Pyke, and S. T. Rachev (Editors), Athens Conference on Applied Probability and Time Series Analysis Volume I: Applied Probability In Honor of J.M. Gani. viii, 424 pages, 1996.

Vol. 115: P.M. Robinson, M. Rosenblatt (Editors), Athens Conference on Applied Probability and Time Series Analysis Volume II: Time Series Analysis In Memory of E.J. Hannan. viii, 448 pages, 1996.

Vol. 116: Genshiro Kitagawa and Will Gersch, Smoothness Priors Analysis of Time Series. x, 261 pages, 1996.

Vol. 117: Paul Glasserman, Karl Sigman, David D. Yao (Editors), Stochastic Networks. xii, 298, 1996.

Vol. 118: Radford M. Neal, Bayesian Learning for Neural Networks. xv, 183, 1996.

Vol. 119: Masanao Aoki, Arthur M. Havenner, Applications of Computer Aided Time Series Modeling. ix, 329 pages, 1997.

Vol. 120: Maia Berkane, Latent Variable Modeling and Applications to Causality. vi, 288 pages, 1997.

Vol. 121: Constantine Gatsonis, James S. Hodges, Robert E. Kass, Robert McCulloch, Peter Rossi, Nozer D. Singpurwalla (Editors), Case Studies in Bayesian Statistics, Volume III. xvi, 487 pages, 1997.

Vol. 122: Timothy G. Gregoire, David R. Brillinger, Peter J. Diggle, Estelle Russek-Cohen, William G. Warren, Russell D. Wolfinger (Editors), Modeling Longitudinal and Spatially Correlated Data. x, 402 pages, 1997.

Vol. 123: D. Y. Lin and T. R. Fleming (Editors), Proceedings of the First Seattle Symposium in Biostatistics: Survival Analysis. xiii, 308 pages, 1997.

Vol. 124: Christine H. Müller, Robust Planning and Analysis of Experiments. x, 234 pages, 1997.

Vol. 125: Valerii V. Fedorov and Peter Hackl, Model-oriented Design of Experiments. viii, 117 pages, 1997.

Vol. 126: Geert Verbeke and Geert Molenberghs, Linear Mixed Models in Practice: A SAS-Oriented Approach. xiii, 306 pages, 1997.

Vol. 127: Harald Niederreiter, Peter Hellekalek, Gerhard Larcher, and Peter Zinterhof (Editors), Monte Carlo and Quasi-Monte Carlo Methods 1996, xii, 448 pages, 1997.

Vol. 128: L. Accardi and C.C. Heyde (Editors), Probability Towards 2000, x, 356 pages, 1998.

Vol. 129: Wolfgang Härdle, Gerard Kerkyacharian, Dominique Picard, and Alexander Tsybakov, Wavelets, Approximation, and Statistical Applications, xvi, 265 pages, 1998.

Vol. 130: Bo-Cheng Wei, Exponential Family Nonlinear Models, ix, 240 pages, 1998.

Vol. 131: Joel L. Horowitz, Semiparametric Methods in Econometrics, ix, 204 pages, 1998.

Vol. 132: Douglas Nychka, Walter W. Piegorsch, and Lawrence H. Cox (Editors), Case Studies in Environmental Statistics, viii, 200 pages, 1998.

Vol. 133: Dipak Dey, Peter Müller, and Debajyoti Sinha (Editors), Practical Nonparametric and Semiparametric Bayesian Statistics, xv, 408 pages, 1998.

Vol. 134: Yu. A. Kutoyants, Statistical Inference For Spatial Poisson Processes, vii, 284 pages, 1998.

Vol. 135: Christian P. Robert, Discretization and MCMC Convergence Assessment, x, 192 pages, 1998.

Vol. 136: Gregory C. Reinsel, Raja P. Velu, Multivariate Reduced-Rank Regression, xiii, 272 pages, 1998.

Vol. 137: V. Seshadri, The Inverse Gaussian Distribution: Statistical Theory and Applications, xi, 360 pages, 1998.

Vol. 138: Peter Hellekalek, Gerhard Larcher (Editors), Random and Quasi-Random Point Sets, xi, 352 pages, 1998.

Vol. 139: Roger B. Nelsen, An Introduction to Copulas, xi, 232 pages, 1999.

Vol. 140: Constantine Gatsonis, Robert E. Kass, Bradley Carlin, Alicia Carriquiry, Andrew Gelman, Isabella Verdinelli, Mike West (Editors), Case Studies in Bayesian Statistics, Volume IV, xvi, 456 pages, 1999.

Vol. 141: Peter Müller, Brani Vidakovic (Editors), Bayesian Inference in Wavelet Based Models, xi, 394 pages, 1999.

Vol. 142: György Terdik, Bilinear Stochastic Models and Related Problems of Nonlinear Time Series Analysis: A Frequency Domain Approach, xi, 258 pages, 1999.

Vol. 143: Russell Barton, Graphical Methods for the Design of Experiments, x, 208 pages, 1999.

Vol. 144: L. Mark Berliner, Douglas Nychka, and Timothy Hoar (Editors), Case Studies in Statistics and the Atmospheric Sciences, x, 208 pages, 1999.

Vol. 145: James H. Matis and Thomas R. Kiffe, Stochastic Population Models, x, 216 pages, 2000.